D0641156

Advances in

ECOLOGICAL RESEARCH

VOLUME 10

Advances in

ECOLOGICAL
RESEARCH

Edited by

A. MACFADYEN

*School of Biological and Environmental Studies, New University of Ulster,
Coleraine, County Londonderry, Northern Ireland*

VOLUME 10

1977

ACADEMIC PRESS

London New York San Francisco

A Subsidiary of Harcourt Brace Jovanovich, Publishers

ACADEMIC PRESS INC. (LONDON) LTD.
24/28 Oval Road
London NW1

United States Edition published by
ACADEMIC PRESS INC.
111 Fifth Avenue
New York, New York 10003

Library of Congress Catalog Card Number: 62–21479
ISBN: 0–12–013910–3

PRINTED IN GREAT BRITAIN BY
T. AND A. CONSTABLE LTD., EDINBURGH

Contributors to Volume 10

PETER CALOW, *Department of Zoology, University of Glasgow, Glasgow G12 8QQ, Scotland.*

JAMES K. JOSLIN, *Department of Biology, George Mason University, Fairfax, Virginia, USA.*

THOMAS F. WATERS, *Depart of Entomology, Fisheries and Wildlife, University of Minnesota, St Paul, Minnesota 55108, USA.*

Preface

The main purpose of volumes in this series is to draw together primary material from currently important fields of ecology and to provide a synthesis for students and specialists from other fields. The range of ecological research topics is now so wide and their state of development so disparate that articles must inevitably differ widely in treatment. This volume's first article by Peter Calow welds together a new look at some classical problems on the borders between ecology and physiology, such as the relative advantages of size versus speed of development and the significance of ageing, with newer approaches through systems theory and cybernetics to physiology and population ecology. It is an important contribution from a most active worker in a field which is currently making rapid progress.

James Joslin's account of homing in small mammals, in contrast, is more concerned with techniques and the interpretation of hard-won field data because the topic is at an earlier stage of development, suffers from a diversity of incompatible techniques and still seeks means by which to evaluate the many rival hypotheses.

The work of the International Biological Programme, although now officially terminated, continues to produce many publications, some of them detailed local studies and others in the nature of syntheses. Almost all of the latter, however, are multi-author volumes, many of which lack the consistency and single-mindedness which can be achieved by an author who examines the meaning and implications of a single theme. Thomas F. Waters' article on secondary production of inland waters is an important contribution of the latter kind because it re-examines the concepts associated with Secondary Production and the many ways in which its measurement can be used to monitor the current state of an ecosystem and assist the ecologist in predicting likely effects of change. There is a useful demonstration of the limited predictive value of the recently fashionable "turnover rate" of P/B ratio and the article also critically reviews the methods of estimating secondary production rates, with emphasis on efficiency and accuracy. This should prove a most valuable guide both to the student in search of understanding and the applied ecologist in need of practical guidance.

May, 1977

A. MACFADYEN

Contents

Ecology, Evolution and Energetics: A Study in Metabolic Adaptation
Peter Calow

Rodent Long Distance Orientation ("Homing")
James K. Joslin

Secondary Production in Inland Waters
THOMAS F. WATERS

Ecology, Evolution and Energetics: A Study in Metabolic Adaptation

PETER CALOW

Department of Zoology, University of Glasgow, Glasgow G12 8QQ, Scotland, U.K.

I. INTRODUCTION

It is traditional to view adaptation in terms of morphology and perhaps even life-cycle but it is often forgotten that though physiologies are generally more homogenous than morphologies they must, nevertheless, be just as in-tune with the environment in which they operate. What follows is concerned with bringing this simple but important

1

idea into focus; with highlighting the relevance of ecology for the physiologist and physiology for the ecologist.

The subject of physiological adaptation is, of course, a very wide one and rather than touching superficially on many aspects I shall concentrate on a few specific problems all concerned more or less directly with the physiology of metabolism. Organisms will be treated as machines, energy transformers, which are designed to work in a certain way and at certain levels of efficiency under particular ecological conditions. In these terms several specific problems present themselves. The first, for example, might be to show that notions of design, teleology, function and the like have a proper place in science; but this is a philosophical question which will be considered only in passing. Instead, it will be understood that the concept of design is not only legitimate but central to biological explanation and within this framework I will consider two general questions: "What design criteria are used in the evolution of metabolic machines?" and "What metabolic strategies are important in the operation of these machines?" From knowing *how* organisms work as energy transformers and *under what circumstances* they transform energy I shall progress to a consideration of *why* they work in the way they do. In the first instance, then, I shall be concerned with the operation of organisms in ecosystems and only in a subsidiary way with the operation of organisms as part of ecosystem metabolism. The latter is, of course, the more usual problem and as "ecological energetics" has been extensively dealt with elsewhere.

II. Historical and Philosophical Perspective

It was Descartes who first introduced the machine analogy to biology as part of a programme concerned with the physical analysis of vital phenomena (Coleman, 1971). His "Bête Machine" soon ran into two serious problems. First, the idea that biology could be reduced to the principles of matter in motion, the science of mechanics, turned out to be one of those scientific *cul-de-sacs* in which we now find abandoned an array of fantastic, though ingenious, models; mechanical models of digestion, nerve action, sensation and even Julien de la Mettrie's "l'Homme Machine" (La Mettrie, 1960). Second, the self-evident fact that machines required an operator to drive them and an intelligence to design them gave impetus rather than discouragement to the arguments of those biologists, the Vitalists as they are called, who believed in an unbridgeable gap between beings and things.

Even in physics, of course, mechanics fell from its pinnacle and thermodynamics took its place for a time as the centre of focus. Under the influence of Priestley, Lavoisier, Atwood, Rosa, and Rubner (Cathcart, 1953) the principles and practices of thermodynamics were also introduced into biology and there they have flourished. Now, for example, it is paradigmatic of bioenergetics that organisms *are* machines, taking in and giving up energy in perfect accord with the laws of thermodynamics, and thereby maintaining themselves at steady-state some distance away from thermodynamic equilibrium.

The idea of design has also taken on a respectable rôle in biology. Darwin's theory succeeded not in reducing design out of biological science but in firmly establishing it as a cornerstone in biological theory. What it got rid of was the mystery behind the concept of design replacing it with a perfectly natural process based on selection, through differential survival, of organisms best fitted to live under particular ecological conditions (Mayr, 1961). Mendel's genes provided a means of design transmission, and molecular biology, by reducing the gene to the chemistry of nucleic acids, has exposed the mechanism behind design transcription, and cybernetic regulation. If the zygote contains the design specifications of the adult in a coded form and if there are feedback mechanisms which constrain development to the prescribed trajectories, then with molecular genetics the mystery has also been taken out of the notions of purposiveness and teleology.

The Cartesian spirit, if not the biological machine as Descartes conceived it, *has* become an established part of modern biology. Quite often, though, questions of design and questions of "mechanics" are dealt with by two different sub-disciplines; evolutionary biology and physiology. The one is concerned with remote causes, general design criteria and fossils without a physiology; the other is concerned with immediate causes and the physiological properties of particular designs. In-between, another group of biologists have become interested in the relationship between the principles of design and the product as manifest under different ecological conditions. This group starts with the assumption that those "machines" occupying particular ecological circumstances now, are the latest products of evolutionary design and that comparative investigations on the design specifications of systems occupying different types of habitats and niches give an insight into why certain specifications have been adopted rather than others. It is always easy, of course, to come up with an answer, the same answer, to this sort of question; one design is adopted (survives) rather than another because it is more appropriate (fitter)—it is obviously more appropriate because it has survived! But this leads at best to triviality and at worst to a rather vicious sort of circularity; fitness = survival = fitness. To see what else is required for a satisfactory explanation, particularly in the context of the "metabolic machine", it is instructive to consider what engineers do in the design and construction of technological systems.

In designing a machine the engineer has three sets of facts: (a) what is required, (b) where it is required (the conditions in which it will operate), (c) what is possible. The general laws of physics set the constraints referred to in (c), by leaving open a plethora of possibilities from which systems might be selected in accord with (a) and (b). The selection of systems from (c) which conform to (a) and (b) gives a very general definition of design. If asked in retrospect how he had come to his final decision the engineer would not reply just that it was the most appropriate design but instead would point to the specific interaction between (a), (b) and (c) which made it the most appropriate.

The biologist possesses similar facts to the engineer, viz: (a^1) a general theory of selection, (b^1) ecological information, (c^1) physiological information, the latter usually making contact with physics. In giving a non-trivial and non-circular answer to the adaptation problem the biologist can step into the boots of the "natural engineer" selecting systems from c^1 in accordance with a^1 and b^1 and comparing the result with the product as it is found to exist. My strategy, then, in line with this general philosophy, will be: (a^{11}) to decide what is required of the metabolic machine (Section III), (c^{11}) to discover what is physiologically possible (Section IV) and (b^{11}) to determine what is appropriate under particular ecological conditions (Sections V–XII).

III. Cybernetic Logic of Life

Selection, the designer, has as its prime requirement the ability of living machines to process energy in order to survive. However, it is not concerned merely with their ability to survive as individuals but with their ability to survive as a design. Energy transformations and the survival of individual organisms are necessary but not sufficient conditions for biological success as judged by selection. Technological machines "metabolize" but do not come under the compass of natural selection and though individuals may survive very well without reproduction (Section XII) their design must necessarily die with them unless they breed. Selection, then, is not primarily interested in "La Bête Machine" as an energy transformer but in "La Bête Machine" as a *transmitter* of genetic information. The logic of life is not based on mechanics nor on thermodynamics but rather on the transmission, reception and storage of messages. "La Bête Machine" is above all a cybernetic machine.

Butler once expressed this genetic view of organisms very neatly when he said that the "hen is the egg's way of making another egg".

The phenotype is thus conceived as a device made by the genome for the survival and propagation of genetic instructions. This genetic reorientation puts phenotypic design and adaptation into perspective; all phenotypes are means to the same end—genetic transmission. In a cybernetic sense they can be considered as transmission machinery and in a chemical sense they are "macro-catalysts" which encourage the precise replication of polynucleotides. Unlike most communication and chemical systems, though, the transmitters and catalysts are the products of their own activity.

A simplified cybernetic model of an asexual organism is illustrated in Fig. 1. Note the two-way flow of information from each genome (M); one channel represents the DNA→RNA→PROTEIN (transcription, translation) sequence, the other represents the replication-reproduction

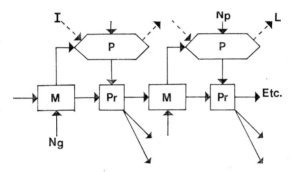

Fig. 1. The cybernetic logic of life. The system is asexual since there is no mingling of gametic information. Sub-systems: M = genome; P = phenotype; Pr = "genetic printer". Information flow is represented by solid lines and energy flow by broken lines. Ng and Np = noise. I = input, L = loss.

process. As already noted there is a catalytic link between the phenotype (P) and the between-individual flow of information; a small amount of information switches on the printing mechanism and thereby triggers the flow of a far larger amount of information. In the main, however, information flows in one way only, from M to P; a feature which is known as the "Central Dogma" of molecular biology and which is designed to protect M from noisy phenotypic nonsense. In effect this means that though changes in the properties and behaviour of P may occur through disturbances in energy input (I) and as somatic mutations (N_p) and although these changes may alter, perhaps even enhance replication, they will not be written into M and in consequence they will not be a feature of the next generation. Lamarckism is not part of the cybernetic logic of organisms. The only changes in design that are transmitted are those brought about by alterations to the genetic system, perhaps by mutation (N_g). Phenotypic characters resulting from alterations of this kind will then come to occupy a greater or lesser part of the future population of the MP-system dependent on how well they favour the replication of M through the catalytic effect of their action on P.

The major difference between the asexual system depicted in Fig. 1 and sexual systems is the possibility, through the intervention of meiosis, and with it crossing over and recombination, for there to be flow of genetic information between the separate lineages constituting a population. This has two consequences. First, by allowing an inter-mixing of alleles sexuality increases the potential variation in the products of reproduction irrespective of mutation. Second, the unit for selection is no longer the genes within individuals but the gene pool within the population. In this case it is meaningful to talk about the fitness of individual genes rather than the fitness of the total genome carried by an asexual lineage. In both situations, though, fitness is measured as the change in frequency of a genetically determined character or set of characters from generation to generation. In this scheme the function of the organism is to carry the genes and to transmit them.

If life has meaning then, if it has some underlying purpose, it must be in the continuing transmission of a message which simply says "transmit me" and in the means for effecting this command. If there is a secret to life, an answer to Schrödinger's question (Schrödinger, 1944), this must be it. Each living organism is merely a transition in this on-going process; a stage between what was and what will be. Selection, of course, operates not on the coded genetic message directly but on its expression in the phenotype. That is, on the phenotype as a transmitter; a rôle that includes the uptake and transformation of matter and energy as well as actual genetic replication itself. By moving the genes from place to place, protecting them against damage and destruction prior to replication, and by providing a suitable "milieu" for their replication, organismic metabolism catalyses the genetic copying mechanism. It is in this sense that we must judge design and organismic efficiency.

IV. What Sort of Efficiencies are There and What Efficiencies are Important?

Any working mechanism requires to be organized and the genetic copying mechanism is no exception. Technological systems are organized by men and are preserved in organization, against the universal tendency for disorganization, by repair. Living organisms, on the other hand, are self-organizing systems which require a continual through-flow of energy to replace spent tissue and to reorganize disorganizing materials. In a very general way it can be said that the positive entropy changes going on within living systems are compensated for (and in some cases more than compensated for) by the import of negative entropy. The thermodynamics of living systems can thus be expressed by Prigogine's (Prigogine, 1955) equation which, in effect, is a restatement of the second law of thermodynamics for open systems — viz.:

$$dS = diS + deS \qquad (1)$$

where: diS is the entropy change within the system, deS is the entropy change by export and dS is the total change within the system. The second law of thermodynamics demands that deS is positive but puts no special constraints on diS. It is therefore possible for diS to be negative and for $|diS| > |deS|$. In this case dS becomes positive and in consequence the entropy changes in open systems may, without violating the second law, be negative. This gives a completely physicalistic answer to what many biologists have thought to be one of life's most special and mysterious properties.

As energy transforming systems living organisms are fundamentally open (see also Bertalanffy, 1968) with an input (I) as food and several outputs (O). These terms, however, are not without ambiguity. In most organisms it is necessary to distinguish between actual and apparent input; the food ingested (I) is the apparent input but only a fraction of this is digested and absorbed (A) and is actually available for metabolism. The difference, $I–A$, is the energy egested (F). Similarly, it is customary to speak of the products of machines as their outputs but in biology some of these products actually form part of the organism, i.e. new protoplasm, and it seems strange to talk about them in the same way. Nevertheless, being end products of "machine activity" they are, in a strict sense "machine outputs" and because they are used by the "machines" I shall refer to them more specifically as useful outputs (UO). Equation 1 demands that $UO < A$, $A–UO$ representing disorganized non-useful (in a biological sense) energy which usually emanates from the body surface as heat. UO can be partitioned further into energy which becomes built into the protoplasm (G) and energy used to form the gametes (Rep). The conversion of I to A and the subsequent partitioning of metabolizable energy is summarized in Fig. 2.

As energy flows through the "metabolic machine" then, it is partitioned into several compartments and, in accordance with Eqn. 1, becomes partially degraded. Therefore, it is natural to start asking questions about the efficiency of the "machine" at diverting energy

along specific routes. In engineering, the word "efficiency" has come to
have a very special meaning, usually referring to the dimensionless
ratio UO/A. But maximizing this ratio is really only one way that a
machine can be judged efficient in the sense of *doing a good job*. One
could, for example, judge efficiency in terms of the speed that A was
converted to UO, or in terms of the constancy of UO in the face of
disturbances in A; for example, by increasing UO/A as A reduces.

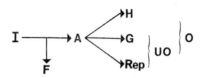

Fig. 2. Partitioning energy in animals. I = input; F = egesta; A = absorbed energy;
H = heat output; G = new protoplasm; *Rep* = reproducta; O = total output; UO =
useful output.

"Biological machines" can, of course, be judged in the same way and
much confusion has arisen in the past through loose use of the term
"efficiency" (see also Slobodkin, 1960; McClendon, 1975). As well, the
problem is confounded in biology because of the many possible ways of
partitioning A and because of the distinction between I and A described
above. In what follows I will use the term "efficiency" in the traditional
sense, as a ratio of output to input (Table I) referring to the rate at
which A can be converted to UO as *"speed of transformation"* and the

TABLE I

Efficiency (output/input) ratios used in the text. Notation from Fig. 2

Efficiency	Symbol	Ratio†
Absorption efficiency*	AE	A/I, $I\text{-}F/I$
Gross growth efficiency	K_1	G/I
Net growth efficiency	K_2	G/A

† Usually expressed as a percentage.
* Sometimes referred to as assimilation efficiency—in physiological terms this is
incorrect (Calow and Fletcher, 1972).

ability of the organism to keep constant UO despite disturbance in A
as *"homeostatic ability"*.
     The efficiency of a machine and the speed of transformation depend
upon both design and environment. There is also a general relationship
between efficiency and the speed of transformation such that maximum
efficiency is incompatible with the maximum rate of transformation and

vice-versa (Odum and Pinkerton, 1955). For a given size it is possible to design a machine which maximizes UO/unit time or a machine which maximizes UO/I (or UO/A) but not both. On the other hand, since "biological machines" incur a maintenance cost even when they are not "productive" the transformation efficiency is likely to fall off even at slow speeds. The relationship between metabolic efficiency and speed of transformation will therefore be of a rising and declining type (Fig. 3). It follows from this that any environmental factor which influences through-flow, e.g. by reducing A, must also influence in a

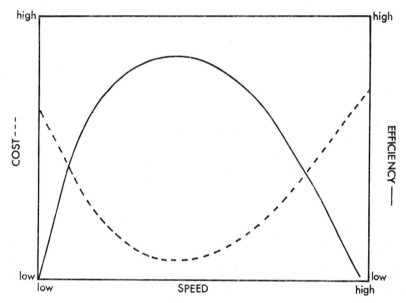

Fig. 3. Theoretical relationship between speed, efficiency and cost of production.

passive way the efficiency ratio UO/A—increasing it to a certain level and reducing it below that level. This is one aspect of homeostatic ability. It is also possible to build machines which are designed to modify UO/A with respect to A such that efficiency can be actively raised when A falls and *vice-versa*. This cybernetic behaviour is another aspect of homeostatic ability. Both aspects will be considered in more detail in Section XIII.

Of course, the most important efficiency for a biological machine is in terms of information not energy (Section III). That is to say, selection judges efficiency in terms of the number of copyable copies that a particular type of organization can generate from a single genetic programme, not in terms of its ability to effect thermodynamic transformations *per se*. As a very rough first approximation, however, fitness must be proportional to the amount of energy put into UO since this will be related, though not directly, to gamete production. Lotka's law of maximum energy which states that "in the struggle for existence, the advantage must go to those organisms whose energy capturing devices are most efficient in directing available energy into channels favourable for the preservation of the species" (Lotka, 1922; p. 147) comes to much the same conclusion. Accordingly we can decide whether biological machines should be designed to maximize efficiency (UO/I or UO/A), to maximize speed of transformation ($UO/$time) or to maximize homeostatic ability. Of course, all three strategies will do. The first essential is that the system is able to obtain energy, but then maximizing the amount passing into UO can either be achieved by increasing $UO/$time for a given efficiency or increasing UO/A (in a phylogenetic or ontogenetic sense) for a given A. There is no warrant, therefore, for the assertion that speed will always be at a premium (c.f. Odum and Pinkerton, 1955) in individual organisms. Whether speed, efficiency or homeostatic ability is selected for will depend on environmental circumstances. Apart from the very general law that UO should be maximized there is no law which specifies how this should be achieved. It is possible, however, that evolution will have tended to shift from speed to efficiency in the course of time since it is likely that those niches filled last would be places where resources were most difficult to obtain (Lotka, 1922). Roughgarden (1971) has also argued that speed, or productivity, is at a premium in situations of "r-selection" whereas efficiency is at a premium in situations of "K-selection". This is because efficiency is needed where competition is keen and resources are in short supply whereas speed is required to fully exploit an uncertain environment.

V. SIZE *VERSUS* FECUNDITY

Not all of UO is used directly in replication, some is diverted to G and so it is interesting to consider to what extent and when increases in G might lead to increases in fitness. In that the development of G will necessarily use up energy and time that might otherwise have been used in *Rep* there is no obvious reason why "bigness" should represent a "good" design feature of an efficient "genetic copier". On the other hand it is indisputable that organisms do grow, some to appreciable sizes, and that in many groups, evolutionary pressure has apparently operated to increase organismic size (Stanley, 1973; Boucot, 1976). With the rise of each new phyletic organization, for example, the maximum size of life has apparently been pushed upwards (Bonner,

1974). How, then, can we reconcile this size *v.* fecundity paradox? To begin with I shall describe some experiments that have been carried out on the evolution of sub-cellular systems. These illustrate, in a particularly clear way, an evolutionary principle which has direct relevance to the size-fecundity problem. Having established the principle I shall then apply it to organismic systems.

A. SPIEGELMAN'S EXPERIMENTS

In a series of very interesting experiments Spiegelman (1971) has asked if, *in-vitro*, it is possible to subject naked genes to artificial selection. Using RNA replicase and a naked RNA of virus origin he was able, by altering culture conditions, to select molecular species of nucleic acid adapted to the test-tube environment in which they were propagated. In one experiment, for example, Spiegelman selected for fast replication and succeeded within just eight generations in dramatically raising the rate of turnover. This character was associated with size reduction; shortening the RNA molecule. Since rapid replication ensures more "molecular offspring" one would suppose that such a mutation should be of considerable selective value and this, of course, was true for the artificial environment. However, the small RNA mutant had become so specialized that when returned to a natural environment it would not reinfect host cells; it had apparently jettisoned the codons which originally permitted the production of coat proteins and the attachment apparatus in the "adult" virus. Only sufficient information to manufacture a replicase was needed for success in the protected test-tube environment but in nature specification of coat and attachment proteins was absolutely essential. The moral to be drawn from these findings is this: In theory the most efficient replicating system is a small template and an enzyme which catalyses duplication; a replicase is, in a sense, the minimum phenotype. But this is a very vulnerable method of reproduction. Other structures and processes must be associated with the replica-replicase system to cushion it from environmental disturbance and thus to catalyse the transmission of genetic information. As is common in evolution there is a conflict of requirements and the result must represent a compromise between rapidity on the one hand and safety on the other. The ecological problem comes in deciding under what conditions the balance tips in favour of one character rather than the other. This problem, "on being the right size", is neither new nor original (Haldane, 1928) but it is not the sort of problem that ecologists have paid much attention to in recent times (Southwood, 1976).

B. HOW TO BECOME BIGGER

There are at least two ways to become bigger (Fig. 4). A genetic change might either raise the growth trajectory by increasing the rate of accumulation of energy in G (a in Fig. 4) or raise the final size reached by lengthening the growing period (b in Fig. 4). In that reproduction usually begins when growth ceases, these changes may on the one hand speed up turnover ("a-type" changes) while on the other delaying it ("b-type" changes). Now, since the spread of a particular gene is like

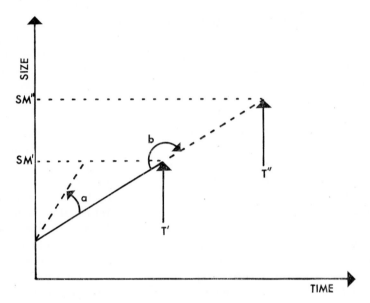

Fig. 4. Two ways to become bigger. The solid line represents the original condition with SM^1 as the size at maturity and T^1 as the age at maturity. Change a involves an increase in growth rate and means that SM^1 is reached more quickly. Change b involves lengthening the life-span with SM^{11} as the new size at maturity and T^{11} the new age at maturity.

the growth of money in a bank account, by compound interest, and since early reproduction is analogous to frequent collecting of interest then "a-type" adaptations have an obvious advantage. Alternatively "b-type" adaptations have no very obvious advantages at all.

Notice, then, that in comparing the relative merits of different sizes there are two separate issues which are not always clearly distinguished: (i) the comparison of contemporaries that have managed to achieve different stages at the same time; (ii) the comparisons of contemporaries who at the same time take different "decisions" in terms of stopping growing and starting to breed.

C. GROWING FASTER

We are not interested in how faster growth is achieved, by speed or efficiency, since this has been dealt with in Section IV. Rather we are interested in how this character might catalyse replication. Already we have noted that by "collecting interest more frequently" an increased developmental rate will be advantageous. However, other factors, like climate, might more often than not prevent this advantage being realized; for example by restricting reproduction to a fixed period. Here, timing and external conditions are more important in determining when an organism becomes mature and reproduces than size is. Temperate, freshwater pulmonates, for example, are annual, breed in spring and die. The resulting young may hatch and grow through to the minimum size needed for reproduction by the onset of winter. However, in Britain breeding is always inhibited until the following spring when temperature rises above a critical value, usually of about 7–8°C (Russell-Hunter, 1961). Therefore, we need to consider how fast growth might influence fitness by means other than speeding up the life-cycle. The general question then, in this section, concerns the relative fitness of differently sized contemporaries.

To begin with it is necessary to make absolutely clear what selection does not take into account in the evolution of size increases. The partitioning of an equal quantity of biomass into a small number of large units invariably results in less catabolic output per unit weight than in the partitioning of the same biomass into a large number of small units (see below). Therefore, size increase at the level of the individual will enable the maintenance of a greater biomass at the level of the population per unit energy flow. This, however, is an effect, not an evolutionary cause of individual increases in size. Though reducing the per gram loss of energy from individuals, size increases raise the absolute loss from the system. In consequence increases in size do not conserve energy for the individual and if this character is to be construed as a component of fitness it can only be done in terms of group selection, an already much disputed topic. Except in the social insects, it is certainly difficult to see how metabolic altruism might operate and in this context it seems particularly dangerous to make comments like, "The problem of evolving species-specific size may be thought in terms of alternative strategies for dividing the amount of energy available by a species" (Schoener and Janzen, 1968; p. 218).

Bigness has three major advantages for the individual. First, larger females tend to produce a larger number of offspring (see Section V, D). Second, bigger organisms tend to be at an advantage in competitive situations involving direct aggression, because, in general, bigger individuals tend to be stronger. Third, and for the same reason, size may confer greater immunity against predation. Bigness may therefore increase the chances of an organism reaching maturity and then increase its reproductive output. These are obvious effects of bigness but there are also more subtle benefits and these involve resource availability, locomotory power and the surface-volume relationship.

Figure 5 illustrates how size might confer a competitive advantage in resource exploitation. The data is from the freshwater limpet, *Ancylus fluviatilis*, an algal feeder which grazes from submerged rock surfaces in fast-flowing lotic habitats and wave-swept littoral regions. Of the algae available for ingestion, different genera have different size ranges as do species within genera and individuals within species.

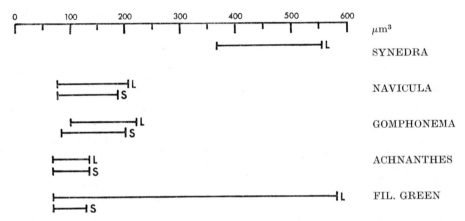

Fig. 5. Size range (volume) of algal cells eaten by large (*L*) and small (*S*) freshwater limpets.

The larger snails can eat algae ranging in volume from *ca.* 75–550 μm^3 whereas the small individuals have a much more restricted range; *ca.* 75–200 μm^3. This means that in times of famine larger individuals have a greater possible choice of food and thus a greater chance of obtaining something to eat. For other examples of this phenomenon see Brooks and Dodson (1965), Weatherly (1966), Galbraith (1967), Hall *et al.* (1971), Emlen (1973), Wilson (1973 and 1974). For a theoretical account, see Wilson (1975).

Turning now to the locomotory effects of size change, Hill (1950) has shown that due to the stresses and strains set up during movement and inherent limits to the strength of biological material, geometrically similar animals should effect similar movements at a rate directly proportional to their linear dimensions. If one animal is 1000 times heavier than another it will be 10 times longer and it will take 10 times longer to make one step. Since, however, the step of the bigger animal will be 10 times longer than that of the smaller animal both will cover the same distance in the same time. The energy expenditure per big step is likely to be less than the expenditure per larger number of small steps. Big animals therefore take longer than small animals to become exhausted and this is the main advantage conferred by size increase on locomotory power. Provided the comparison is restricted to geometrically similar bodies, then, size is found to influence the efficiency not the speed of locomotion. Since intra-specific competition and the "struggle for existence" occur between individuals of similar morphological design the constraint of similitude is relevant. In consequence the apparently contradictory finding of Bonner (1974), that speed increases with size, is not very interesting as a possible evolutionary source of size change since his comparison was based on inter-phyletic differences and therefore on drastic changes in shape as well as size. If Hill's argument is accepted, inter-phyletic increases in speed depend more on structural modifications than on size changes.

Finally, to consider the effect of size on the relationship between body surface area and volume. Volume and mass increase as the cube of the length whereas area increases as the square. Therefore, the surface-area to volume ($S : V$) ratio reduces with volume. Once again this strictly applies to geometrically similar bodies only. Since body surface is an important point of contact with the external environment the $S : V$ ratio is an important physiological parameter and changes in it are likely to have important physiological consequences. Most processes involving interchange between internal and external environment, for example, are related directly to surface. The physiological consequences of alterations in the $S : V$ ratio, of course, will vary from group to group and a useful distinction in this respect is between homeotherms and poikilotherms. In the former the $S : V$ ratio is of importance for temperature regulation, in the latter for the conservation of matter, and in terrestrial habitats, particularly water. Once again it should be remembered that increase in size reduces the $S : V$ ratio and thus the per gram output of material and energy. They are therefore important in maintaining the constancy of the "milieu" on a per gram basis. Alternatively, as McNab (1971) argues, size increases necessarily lead to an increased absolute loss and cannot be considered as a device for reducing the total demands or through-flow of energy and/or matter of the system. Given an adequate supply of energy and materials, however, reductions on a per gram (or per volume) basis become important when the proper functioning of the "metabolic machine" depends on the maintenance of a critical tissue concentration of some property. This is true of tissue water content and, in homeotherms, of tissue heat levels. Therefore, though McNab (1971) is correct when he points out that size increases may be of selective disadvantage to polar homeotherms facing food shortage they may nevertheless contribute to the per gram maintenance of temperature under the thermal stress of a polar environment and thus be of positive advantage.

Heat loss from the body surface of homeotherms depends on a number of factors, viz.: (a) the temperature gradient between internal and external environment; (b) the extent to which the internal environment makes contact with the external ($S : V$ ratio); (c) the form and extent of insulation (e.g. fur, feathers and lipid). Since there are no dramatic changes in body temperature between mammals and birds living in hot and cold places, cold-adapted homeotherms must either increase their insulation or reduce their surface area; the converse must be true of hot-adapted individuals. Scholander (1955) argues that insulation rather than size changes are important since a 10-fold increase in the temperature gradient, which can occur from tropics to poles, would result in a 10-fold cooling effect. This could be accounted for by just a few centimetres extra fur or fat but would require a 10-fold reduction in surface and therefore a 1000-fold increase in mass! The importance of insulation as opposed to size is therefore beyond dispute but it must still be admitted that any benefit, however slight, accruing to the per gram regulation of temperature from $S : V$ adjustments will be selected for. Provided comparisons are restricted to within species, Bergmann's rule, the tropico-polar tendency for organisms to increase in size, still seems to be a satisfactory statement of a real eco-geographic trend (Mayr, 1956).

Temperature regulation is unlikely to be very important in the size specification of poikilotherms. Here the apparent conformity with Bergmann's rule seems to be of phenotypic rather than genotypic origin (Bertalanffy, 1960). Unlike the mammals, however, most invertebrates have poor cutaneous mechanisms for conserving body water. One of the major avenues for control here, therefore, might be through the reduced $S : V$ ratio. Schoener and Janzen (1968) have, in fact, found a significant negative correlation between insect size and environmental humidity and Nevo (1973), working on the adaptive size of cricket frogs in North America, also found a negative correlation between body length and humidity. Here a combination of three humidity variables explained 60% of the variation in size data.

To summarize: Increased growth rate may contribute to fitness by speeding up the life-cycle and/or by exploiting the advantage of size in organisms of the same age. In general, bigger organisms have better chances of surviving through to reproduction and then of producing more gametes. This is particularly true of homeotherms in cold climates and poikilotherms in dry habitats. There is, however, a price to be paid for size increases. As the size of the individuals in a population increases, the size of the population usually decreases. Thus organisms which (in an evolutionary sense) increase in size more rapidly should become extinct sooner. Hallam (1975) has shown this to be the case for Jurassic bivalves and ammonites.

D. GROWING FOR LONGER

Sometimes, as in barnacles (Barnes, 1962) and fish (Gerking, 1959), reproduction affects growth momentarily; sometimes, as in insects and the higher vertebrates, reproduction only occurs when growth has ceased. In any event energy diverted to Rep is denied to G and *vice-versa*. In some cases the decision between growth and reproduction is a straight forward *"yes-no"* one whereas in others it is a "yes, if certain conditions pertain" one. To begin with we are interested in the most clear-cut decision; to reproduce and stop growing or to grow and not reproduce. In a later section (Section XII) we shall consider the possibility of ontogenetic adaptability in the reproductive strategy.

Consider two individuals both approaching T^1 (in Fig. 4) along the same growth trajectory. One "decides" to breed and stops growing, the other "decides" to continue growing and thereby puts off breeding to some future date, T^{11}. Which is fitter? The compound-interest law demands that, all other things being equal, the fittest individual is the one which breeds first. Furthermore, it is unlikely, in this case, that the conclusion would be influenced by the fact that bigger individuals tend to have a greater reproductive output, for here we are not comparing contemporaries and the gamete production of the older and larger individual not only needs to be greater than the smaller individual but also needs to "make up for lost time". To do this, reproductive output would need to be an exponential function of age and thus size. Consider, for example, another hypothetical comparison between two organisms; one grows 10 size units/unit time and gives rise to 10 offspring after one unit of time and the other grows at the same rate but delays breeding until it is 1000 units big. To compensate, the latter must produce not 10 times more offspring but 10^9 times more offspring! Even if gonads grew at a greater rate than the body as a whole there must be a mechanical limit to expansion. Usually, however, the relationship between egg output and body mass is no better than linear. For example in the arrow worm, *Sagitta elegans*, the maximum number of eggs per female is proportional to her length raised to the power of 2·5 which is in negative allometry with respect to weight

(McLaren, 1966) and a similar relationship occurs in the Lycosidae (Peterson, 1950) and *Gammarus* (Kurtén, 1953). In fish (Gerking, 1959) salamanders (Salthe, 1969) and reptiles (Tinkle *et al.*, 1970) there appears to be an isometric relationship between egg output and body weight, and in the domestic fowl egg numbers are proportional to body surface (Brody, 1945).

The generality of the conclusion that there is a negative correlation between size and fitness, in terms of reproductive potential, has been underlined by the analytical work of Lewontin on colonizing species (Lewontin, 1965).

Of course, evolutionary success is not simply judged in terms of potential reproductive output but in terms of the actual production of viable progeny. If, for example, reproduction puts the parent at risk, and larger individuals survive better from T^1 to T^{11} than smaller individuals then the gain in terms of *"copyable* genetic copies" will ultimately switch in favour of a growth rather than a reproductive strategy. Indeed in this situation the parent need not grow through T^1 to T^{11} and may even shrink without affecting the wisdom of the decision to delay breeding.

There can be no doubt that reproduction, by weakening the parent, making it more conspicuous or subjecting it to metabolic stress (Section XII), does involve considerable risks. Tinkle (1969), for example, plotted mean annual survivorship against fecundity per season for 13 lizard species and found an inverse relationship and Loschiavo (1968) has found much the same relationship for the beetle, *Trogoderma parabile*. We have already discussed the relative merits of bigness over smallness for particular ecological circumstances and the same arguments apply here. If, for example, a dry spell is interpolated between T^1 and T^{11} and the species is a terrestrial poikilotherm, then the environment will favour retention of the adult at the expense of the progeny. Similarly, if conditions are such that competition is keen, large parents are likely to fare better than small progeny. This is particularly interesting in terms of current ecological theory and the notions of "r" and "K" selection since in a stable environment, when populations are at carrying capacity and competition is keen, i.e. towards the "K" end of the selection spectrum, size rather than fecundity is likely to be at a premium. Conversely, in unstable conditions where populations are below carrying capacity, i.e. towards the "r" end of the spectrum, fecundity rather than size will be at a premium. In line with this prediction mean organismic size of the community components, both plant and animal, often rises during succession (Odum, 1971).

To summarize: It is always better to breed rather than grow unless breeding endangers the adult and the adult has better chances of surviving than smaller progeny. The relative merits of bigness *versus* smallness can be considered in terms of hot *versus* cold, dry *versus* wet, predator rich *versus* predator poor and *"r" versus "K"* environments. Low temperatures, dry conditions, high predation and keen competition will favour an extended growth rather than precocious breeding strategy. It is clear that these forces may sometimes work in opposition. Then the character adopted will represent some sort of compromise. This underlines the dangers of thinking in terms of one dichotomy only; e.g. *"r" versus "K"* selection.

VI. Adiposity *versus* Fecundity

Metabolic compartment G can itself be divided into several sub-compartments and this can be effected in a number of ways. G could, for example, be divided anatomically into organs and tissues or bio-chemically into molecules belonging to similar chemical families; fats, proteins, carbohydrates etc. Metabolically, though, it is useful to distinguish between two sub-compartments; the lean body mass and that compartment which often consists of adipose tissue and which is involved in energy storage. Since selection will clearly operate to allocate energy between compartments in such a way as to optimize fitness it is interesting to consider what part each might play.

In these terms it is easy to see that by providing an essential framework for metabolism the lean body mass catalyses more growth (G is autocalytic) and ultimately reproduction (Section V). On the other hand building up the metabolically inert storage compartment effectively slows down maturation and makes no apparent contribution to progeny production. In Slobodkin's words: "there is a selective advantage in increasing fecundity but not adiposity" (Slobodkin, 1962; p. 71). Given a resource limiting biosphere, excess energy should be converted either directly to offspring or to materials which will ultimately promote, not hinder, their production. Significantly, Slobodkin and Richman (1961) found that the potential energy per gram of animal tissues from 17 species, covering six phyla, was skewed towards the lower limit of the possible range of biological materials. This implies that fat, with a high energy value (9·0 kcal/g; 37·8 kJ), is a minor component of animal biomass.

Like all generalizations, that of Slobodkin is in need of some elaboration and correction. First, adipose tissue sometimes has a direct functional effect, as in the insulation of homeotherms (Section V) or the buoyancy of plankters and may thereby contribute directly to fitness. Interestingly, passively floating planktonic animals do tend to

have higher energy values than their actively swimming counterparts
(Wissing *et al.*, 1973). Second, storage itself has an adaptive value as
investment against periods of resource limitation that are interposed
between birth and maturity. These may be self-induced or environ-
mentally mediated. For example, energy values rise as a result of
intense storage prior to pupation in insects (Slobodkin, 1962), migration
in birds (Odum *et al.*, 1963), whales (Brodie, 1975) and termites (Wiegert
and Coleman, 1970) and the onset of seasonally poor, usually winter,
conditions (e.g. Schindler *et al.*, 1971). Third and finally, it might be
anticipated that if the adult can withstand hardship better than its
young and if there is a continual danger of feeding disturbances then
it will always be advantageous for the adult to store some energy, at
the expense of fecundity, as an insurance policy. Paradoxically, this
strategy should be more prevalent in inclement, energy-limiting
environments than in stable, energy-rich situations. The Platyhelminthes provide an almost unique opportunity to test this hypothesis
because this phylum contains a variety of species of common basic
organization but with divergent life-styles and ecologies. Species range
from free-living predators whose ecology is dominated by resource
limitation and intense competition (Reynoldson, 1966) to entosymbiotes
which live in energy-rich situations, with ectosymbiotes occupying an
intermediate position between these trophic extremes. Figure 6 shows
that the distribution of energy values in this group, as expected,
ranges from a high level in the free living species, through intermediate
levels in the ectosymbiotes to a low level in the entosymbiotes (Calow
and Jennings, 1974). The cestode values were high for entosymbiotes
because the tissue samples used in calorimetry contained gravid proglottids. These are egg sacs and are therefore rich in albuminous lipids.
True adult tissue in the cestodes had a low energy value (5·5 kcal
(23·1 kJ)/g ash-free dry weight as opposed to a value of almost 6·0 kcal
(25·2 kJ) for gravid proglottids) in keeping with other entosymbiotic
species (Calow and Jennings, 1974).

Clearly, storage strategies are products of selection and evolve in
response to the challenge presented by the trophic conditions of the
environment. Two conditions lead to "adiposity": (1) predictable
periods of trophic hardship, and (2) unpredictable but probable periods
of trophic hardship. The same general principles apply to micro-
organisms (Parnas and Cohen, 1976). Arguing from the fact that
condition 2 is likely to be characteristic of situations of intense competition Jennings and Calow (1975) have suggested that "adiposity"
and thus "caloricity" is likely to be a product of "K" rather than "r"
selection. Alternatively, reasoning from the fact that an "r" selected
species needs to form as many offspring as possible as soon as possible,

Derickson (1976) has suggested that "*r*" selected lizards, moving into or out of predictable periods of food shortage (condition 1), should contain more fat (for the rapid production of young) than "*K*" selected species. "Adiposity" might therefore be a result of either "*r*" or "*K*"

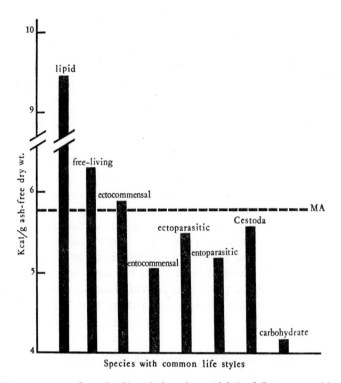

Fig. 6. Mean energy values (kcal/g ash-free dry weight) of flatworms with common life-styles. The values for lipid carbohydrate and *MA* (the mean value for whole animals from other phyla) are from Cummins and Wuychek (1971) (with permission from Calow and Jennings).

selection and in this context these concepts are not very useful. More useful are the notions that storage strategies should be adaptive and that the conditions listed above are the ones which are most likely to provoke adiposity and thus the accumulation of more joules per gram of tissue.

VII. Egg Size *versus* Numbers

The "decision" between size, adiposity and fecundity must also be taken at the stage of gamete formation. A female can either produce a large number of small eggs with low calorific value or a small number

B

of large eggs with high calorific value. Only rarely, in very rich habitats, is it possible to produce a large number of large yolky eggs (Jennings and Calow, 1975).

If a female has a limited energy supply it is clear that she must compromise between egg size and egg numbers. Since fitness depends on the number of viable offspring, selection will tend, in the first instance, to favour those species which partition Rep (Fig. 2) into the most parts; only a small amount of energy is required for information transmission. However, since larger eggs usually result in larger and more fully developed offspring these will be at a premium whenever size at birth is correlated with a better chance of reaching maturity. At least part of the solution to the "egg size *versus* numbers problem" depends upon the relative merits of big *versus* small "hatchlings" and can therefore be decided in much the same terms as already discussed in Section V, C. In general the balance is likely to tip in favour of size rather than numbers as the conditions of "larval" life become harder. Gastropods living in freshwater habitats, for example, produce about $10-10^3$ eggs/individual/lifetime whereas marine species produce about 10^6 eggs/individual/lifetime. This difference is correlated with the fact that as gastropods have invaded the more inclement freshwater environment, free living larval stages have been retained increasingly in the protective envelope of the egg. It has been necessary, therefore, to increase the space and provisions provided for the developing embryo and thus to concentrate on egg size rather than numbers (Russell-Hunter, 1970).

VIII. Trophic Level and Efficiency

Alexander (1975) makes the point that there is an intricate integration of parts in the mechanical systems of the bodies of animals which must be maintained in evolution. This is equally true for the metabolic systems. Here the principle of integration is best illustrated by the fact that reductions in the efficiencies of some metabolic processes, consequent on specific adaptations, usually evoke compensatory modifications in the evolution of other processes so that the overall efficiency of conversion of food to G and Rep is maintained. I illustrate this principle in this and the following section. Then, in Section X, I consider a situation where the principle apparently fails.

In a review of the energetics of 14 species of aquatic poikilotherm Welch (1968) found that there was a clear negative relationship between K_2 (Table I) and absorption efficiency (A/I). In general high absorption efficiencies were associated with carnivory and low efficiencies with herbivory; a feature consonant with the finding that proteins are easier to digest than high molecular weight polysaccharides. Similarly, low net growth efficiencies (G/A) were associated with carnivory and high net efficiencies with herbivory; probably reflecting the relative ease with which vegetable as opposed to animal food is obtained and eaten.

It is easy to see that the effects of these relationships will be to standardize herbivore and carnivore efficiencies in terms of K_1 (Table I). It is not easy, however, to distinguish cause from effect in this situation nor even to ascribe these relationships to some positive selection force. Given that selection is likely to raise the level of both the growth efficiency and the absorption efficiency it is necessary to ask: "Did the low growth efficiencies attendant on active predation lead to the evolution of high absorption efficiencies?" or "Did the low absorption efficiencies attendant on herbivory lead to modifications in conversion efficiency?" or "Is it just a matter of chance that these two features were related at all?" It is extremely difficult to give any final answer to questions of this kind but since we do know that selection may modify the activity of digestive enzymes (Hochaka and Somero, 1973; Calow and Calow, 1975) and may also alter the size specific level of metabolism (Section IX) it seems plausible that the condition noted above was brought about by a positive act of selection. In the absence of facts to the contrary it seems most reasonable to expect that this occurred by a modification of both the absorption (A/I) and the conversion efficiencies (G/I and G/A).

IX. METABOLISM OF MUCUS PRODUCERS

Not all the non-respired portion of absorbed energy is used in growth and reproduction; excretions and secretions may also result in energy losses. These usually represent an insignificant fraction of the input energy and can safely be ignored but in a few instances secretions do become a predominant feature of metabolism. Teal (1957), for example, found that as much as 70% I may be lost as mucus in flatworms and I have found that a smaller but nevertheless significant fraction (approx. 20% I) is lost in the same way by freshwater gastropods (Calow, 1974a). Mucus secretions are, of course, needed by both the flatworms and the snails for movement. Consequently, these groups seem to be committed to a metabolic strategy which is unusually expensive in moving the genetic programme from place to place. This being the case it seems reasonable to expect that compensatory economies might have been made elsewhere in the metabolism of mucus producers. One possibility, for example, might be through reductions in the cost of maintenance, a device apparently used by food-limited spiders (Anderson, 1970).

To test the above hypothesis I have calculated regression lines for
the relationship between the logarithm of respiratory rate (Kcal/indi-
vidual/hr) and the logarithm of body size (grams, shell-free, wet weight)
for four species of triclad turbellarian and 18 species of freshwater
gastropod snails and have compared this with Hemmingsen's standard
regression line for metazoan poikilotherms at 20°C (Hemmingsen, 1960).
Data for the flatworms are from the paper of Whitney (1942) and were
corrected from 14·5°C to 20°C using Krogh's standard correction curve
(Krogh, 1916). Data for the snails are from the paper of Berg and
Ockelman (1959) and include information on both pulmonates and
prosobranchs. No temperature corrections were needed here. It is
difficult to say whether "standard" ("maintenance") metabolism as
such was measured by these workers. However, since departures from
"maintenance" are all positive the resulting imprecision actually
strengthens the test. This is because if any difference is found between
Hemmingsen's prediction and the actual data, it will mean that there
must be an equal and probably greater difference between the prediction
and maintenance metabolism.

The double logarithmic plots are presented in Fig. 7a and b, the
regression equations are listed in Table II, and several null hypotheses

TABLE II

Regression equations relating Log respiratory rate (Y) to Log shell-free, wet weight (X)

Group	Equation	No. replicates	No. species
Standard	$Y = -3·161 + 0·751X$	—	—
Flatworms	$Y = -3·601 (\pm 0·158) + 0·740 (\pm 0·066)X$ $r = 0·936\ S^2_{yx} = 0·010$	20	4
Snails	$Y = -3·130 (\pm 0·105) + 0·948 (\pm 0·046)X$ $r = 0·948\ S^2_{yx} = 0·018$	22	18

Y = Kcal/individual/hr; X = grams; r = correlation coefficient; S^2_{yx} = mean
square deviations from the regression line.

are tested against parameters in the regression equations in Table III.
The regression coefficients for each species were significantly different
from zero indicating a significant linear relationship between dependent
and independent variables. The slope of the "flatworm line" was not
significantly different from the "standard line" but the level of metabolism, as judged by the "a coefficient", as expected was significantly
depressed (by 0·440 log decades or 2·75 times). The snail data were
less conclusive. Though the "a coefficient" of the regression line was

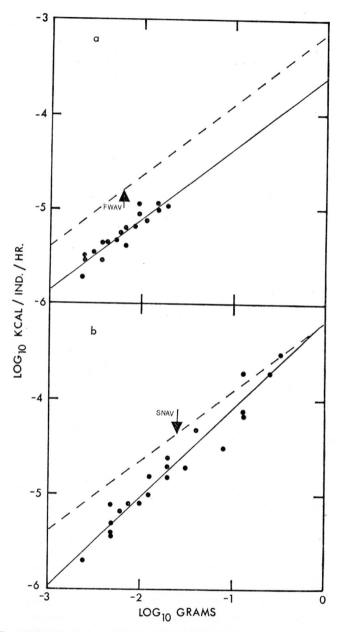

FIG. 7. The relationship between the logarithm of respiration (kcal/hr) and the logarithm of body size (gram, shell-free wet weight) in flatworms (a) and snails (b). The broken line is Hemmingsen's standard regression for poikilotherms. $FWAV$ = average sized flatworm. $SNAV$ = average sized snail.

not significantly different from the standard line the slope of the former was significantly higher than the standard value of 0·75. This means that the "snail line" diverges from the standard and that whereas large snails have rates equivalent to other poikilotherms small snails have reduced respiratory rates (Fig. 7b). Average sized snails (i.e. in terms of the species considered by Berg and Ockelman) have

TABLE III

Testing the parameters in the regression equation in Table II
against several null hypotheses

Null hypothesis	Test*	t	P
$B = 0$ $B = 0.751$	$\left\| \dfrac{b-B}{S_{yx}/\sqrt{(x-\bar{x})^2}} \right\|$	$\begin{cases} FW = 11\cdot252 \\ SN = 20\cdot750 \\ FW = 0\cdot166 \\ SN = 4\cdot520 \end{cases}$	$< 0\cdot001$ $< 0\cdot001$ $> 0\cdot100$ $< 0\cdot001$
$A = -3\cdot161$	$\left\| \dfrac{a-A}{S_{yx}\sqrt{1+1/N+\bar{x}^2/\Sigma\,(x-\bar{x}^2)}} \right\|$	$\begin{matrix} FW = 2\cdot320 \\ SN = 0\cdot241 \end{matrix}$	$< 0\cdot050$ $> 0\cdot100$
$\left.\begin{matrix} AV_{FW} = 4\cdot660 \\ AV_{SN} = 4\cdot340 \end{matrix}\right\}$	$\left\| \dfrac{av-AV}{S_{yx}} \right\|$	$\begin{cases} FW = 4\cdot800 \\ SN = 2\cdot241 \end{cases}$	$< 0\cdot001$ $< 0\cdot050$

B = hypothetical regression coefficient; b = actual regression coefficient from Table II; A = hypothetical intercept on Y axis for $X = 0$; a = actual intercept on Y axis for $X = 0$, from Table II; AV = respiratory rate (Kcal/individual/hour) for an average sized organism as predicted by Hemmingsen's standard curve (see text); av = actual respiratory rate of an average sized organism as obtained from the regression equations in Table II; x, $\bar{x}$ etc. = size of animals (shell-free, wet weight in grams); N = number of observations; S^2_{yx} = mean square of deviations from the regression line; FW = flatworm; SN = snail; t = Student's critical value for significance with $N-2$ degrees of freedom; P = probability of rejecting the null hypothesis when it is correct.
*see Snedecor (1966).

a respiratory rate significantly less than other poikilotherms of the same size (by 0·300 log decades or approx. 2 times). In general, therefore, small snails ($< 0\cdot05$ g) conform to expectations whereas big snails do not. Why this should be so is not yet clear though one possibility is that larger snails produce proportionately less mucus. This certainly seems to be the case for larger terrestrial species (Richardson, 1975).

Before leaving the subject of mucus secretions it is also important to note that this process may represent an investment of energy rather than a complete loss. For example, the mucus trails of flatworms may be used to trap prey (Jennings, 1957) and snail trails seem to promote the growth of bacteria which can themselves be used as food by some snails (Calow, 1974b). Though these effects are difficult to quantify it is quite clear that they are likely to have an important catalytic effect on replication which will offset, to some extent, the energy cost of mucus production.

X. POIKILOTHERMY *VERSUS* HOMEOTHERMY

"Are poikilotherms more efficient energy transformers than homeotherms?" and if so, "how could homeothermy have evolved?" These questions have been tackled by both ecologists and physiologists alike and yet both groups have come to substantially different conclusions. The ecological school, including Engelman (1966), Golley (1968) and McNeill and Lawton (1970) has argued that homeothermy depends upon thermogenesis and that this must lower the efficiency by which ingesta is converted to protoplasm. Alternatively, the physiological school, from Rubner (1924) onwards (Rahn, 1940; Brody, 1945; Kleiber, 1961), has argued that since the decisive point in the development of homeothermy is not the *level* of heat production but rather the regulation of its loss then there are no *a priori* grounds for assuming that homeothermy should be any less efficient than poikilothermy. The final arbiter in a controversy of this kind, of course, must be the facts but the difficulty here has been in making valid comparisons. It is well known, for example, that growth efficiency alters with age, size and environment in both homeotherms and poikilotherms so that comparison is only valid under standard conditions.

One possible way of comparing efficiencies is to use general parameters representing energy input and output rates under fixed conditions. Gross growth efficiency (K_1) is given by $G/I \times 100$ and this can be rewritten as:

$$K_1 = (P - B/I)\,100 \tag{2}$$

where: P = partial efficiency (net energy available for production per unit quantity of ration); I = food ingested; B = basal or fasting catabolism. This equation has been used frequently by Kleiber (1961) who points out that the ratio I/B, which he calls *relative food capacity*, is an important measure of efficiency. This is because as I/B increases in value more food is ingested per unit metabolism and, therefore, the conversion of energy to biomass becomes more efficient. General estimates of I and B per unit "metabolic size" are given by:

$$I = i/W^{0.75}T \tag{3}$$

$$\text{and } B = r/W^{0.75}T \tag{4}$$

where: i = energy ingested (Kcal) per unit time, T (days), by an organism weighing W (grams) wet weight; r = respiratory heat loss (Kcal) per unit time, T (days), by a resting organism. $W^{0.75}$ is used as a measure of "metabolic size" since the metabolic processes are usually related in a direct, linear fashion to this function of body weight.

In a variety of domestic homeotherms Kleiber (1961) found that I ranged over a short interval between 0·253 and 0·452 (average 0·351) and that I/B ranged from 4·2 to 5·7 (average 4·9). Table IV calculates I for a variety of carnivorous fish; i being corrected to 20°C using Krogh's curve (Krogh, 1916). These data were chosen because i was based on

TABLE IV

Calculation of I for six species of fish. See text for explanation

Authority	Species	i†	$W^{0.75}$‡	T§	I§§
Warren and Davis (1967)	*Cichlasoma bimaculatum*	2·62	2·34	14	0·080
	Cottus perplexus	5·50	1·22	55	0·082
	Salmo clarki	8·71	2·64	77	0·194
Pandian (1967)	*Megalops cyprinoides*	1·95	18·84	1	0·104
Pentelow (1939)	*Salmo trutta*	60·39	15·92	21	0·181
Johnson (1966)	*Esox lucius*	125·00	30·41	27	0·152
Average					0·132

† Kcal, adjusted to 20°C using Krogh's curve (Krogh, 1916).
‡ Fresh weight (g).
§ Days.
§§ = $i/W^{0.75}T$.

careful measurement from animals kept on a *non-restricted* diet. Here I ranges over a short interval from 0·080 to 0·194 with an average of 0·132. Using Hemmingsen's data on the relationship between standard metabolic rate and body size (Hemmingsen, 1960; p. 18) B for poikilotherms at 20°C is found to be 0·017. Therefore, the average I/B ratio for fish is 7·8 and this means that they are potentially 1·6 times more efficient in the conversion of food to protoplasm than homeotherms.

It is possible, of course, that not all the potential rise in efficiency expressed by the I/B ratio is realized. This would be the case if there were differences in absorption efficiencies between homeotherms and poikilotherms. In the mid-sixties such differences were thought to exist with Engelman, for example, expressing his belief that poikilotherm absorption efficiencies never rise above 30% whereas homeotherm efficiencies were close to 70% (Engelman, 1966). The implications were that compensatory adjustments in homeotherms are being made at the level of digestion and absorption. However, we now know that the absorption efficiencies of cold-blooded invertebrates may equal if not exceed those of homeotherms both in predatory (Lawton, 1970) and herbivorous (Calow, 1975a) species so the principle of adjustment (Section VIII) does not seem to apply here. There must be an actual difference in the growth efficiencies of homeotherms and poikilotherms even in terms of absorbed energy.

There seems to be at least *prima facie* evidence, therefore, for the claim that a homeotherm is not simply "a poikilotherm with a particular hypothalamic function superimposed" (Adolph, 1951). That is, there seem to be real metabolic differences between these two groups which are probably related to the "speed-efficiency principle" (Section IV). Poikilotherms *are* more efficient than homeotherms in terms of energy transformations but this need not provide any problems with regard to the evolution of homeothermy (cf. Engelman, 1966; p. 105). As we have already seen (Section III) selection is not interested in conversion efficiencies *per se*, but in the relative abilities of different organisms to obtain energy, and then in the rate at which they convert it to viable offspring. The cost of homeothermy in terms of energy is clearly offset by the extent to which constant internal temperatures and high activity levels, irrespective of external conditions, catalyse the capture of energy, the survival of the adult and the subsequent survival of offspring.

XI. METABOLIC ADAPTABILITY

There can be little doubt that it is important for living things to reach particular stages in their development in prescribed times (Section V). Those organisms which can keep to the developmental schedule despite disturbances must be at an advantage over their less adaptable contemporaries. It is reasonable to expect, therefore, that compensatory mechanisms should have evolved to resist those disturbances that tend to deflect organisms from their "desired" developmental trajectories. Of course, given a severe enough and sufficiently prolonged disturbance even the most efficiently controlled organism will die. Hence, physiological homeostasis should be viewed as a temporary solution to environmental disturbance.

One important and common form of perturbation occurs through reductions in the quality and quantity of the food supply and I shall concentrate particularly on this problem. That disturbances in food supply do elicit metabolic compensation is amply demonstrated by the fact that growth efficiencies often increase with reducing rations (Paloheimo and Dickie, 1965, 1966; Kerr, 1971; Kelso, 1972), that growth rate is not directly proportional to food supply (Hubbell, 1971; Calow, 1973a) and that growth "spurts" and "tracking" phenomena often occur after periods of partial or complete growth suspension (Osbourne and Mendel, 1915; Clarke and Smith, 1938; Wilson and Osbourn, 1960; Calow, 1973a). These, of course, are whole-organism manifestations of metabolic regulation and must themselves be dependent on the compensatory activities of physiological sub-systems. How the latter respond to disturbance depends on what form the disturbance takes and on the "life-style" of the organism being disturbed. Predators, for example, which on average receive more frequent disturbances in food supply than herbivores (Hairston *et al.*, 1960), and which exploit a mobile rather than static food source, should in these terms respond more efficiently to starvation than either herbivores or detrivores. It is interesting to find, therefore, that weight loss during starvation does appear to occur at a lower rate in predatory as compared with herbivorous fish (Ivlev, 1961). In this way explaining physiological adaptability as an evolutionary adaptation develops a strong ecological bias. Information on this subject is widely scattered but is very extensive and so my aim here will not be to present a comprehensive review but rather a critical synthesis which reflects in a large part my own specific interests. For convenience I shall distinguish between two distinct, but nevertheless related forms of food supply disturbance; that based on the level of food availability and that based on the quality of the available food.

A. REDUCTIONS IN LEVEL

In principle it is possible to have at least two extreme levels of food supply; the one providing all the food that is required; the other providing no food at all. In practice it is unlikely that either of these extremes will be very common in nature and the more usual situation is likely to be one in which animals experience undernutrition. Here periods of famine will be punctuated by periods when food is more or less readily available. Even so-called continuous feeders will experience a condition of this sort but here the interval between contacts with food will be short.

Undernutrition poses two problems. First, "what should be done when there is no food available?" and second, "what should be done when contact is again made with food?"

a. *Response during famine*

When there is no food immediately available the most obvious response would seem to be to look for more food. However, in that searching involves movement (increased activity in immobile forms, e.g. filtering rate) it also involves an energy cost (see below) and this must be balanced against returns (immediately in terms of energy, ultimately in terms of fitness) before a straightforward decision to "move on" and look for more food can be entertained. It seems likely that the strategy adopted will depend on two factors: (a) the form and extent of the feeder's endogenous energy stores, and (b) the probability of it finding food for a given effort as opposed to it finding food without effort. The latter will depend, in turn on the mobility of the food. If, for example, both food and feeder are immobile the chances of them coming into contact will be zero; as the activity levels of each increase so should the probability of one coming into contact with the other. Herbivorous and detrivorous snails, for example, which contain only short-term glycogen reserves and which feed on stationary food become more active upon starvation (Calow, 1974a; Townsend, 1975) whereas predacious dragonfly larvae which contain long-term, fatty reserves and which feed on mobile food can afford to "sit and wait" for replenishment (Etienne, 1972). In the Lycosidae, *Filistata hibernalis* is a web builder and this increases its chances of finding food without effort whereas *Lycosa lenta* is a related species which does not build a web. Upon starvation the latter becomes more active than the former (Anderson, 1974).

Another group on which it has been possible to examine the influence of the mobility of the food on the searching strategy of the feeder is the Turbellaria. Here, *Dendrocoelum lacteum* is an active predator capable of taking active prey (like live *Asellus*) whereas *Polycelis tenuis* scavenges on dead or dying prey which are immobile (Reynoldson, 1966). It seems reasonable to expect, therefore, that the scavenger should remain active during starvation whereas the predator might become quiescent. To test this hypothesis I measured the rate of movement of groups of worms, each of roughly the same size (extended body length 15 mm *D. lacteum*, 10 mm *P. tenuis*) at regular intervals over a one month starvation interval. Worms were allowed to traverse filter paper cones for 3 hrs. in 250 cm³ crucibles containing filtered lake water. After the experiment, activated charcoal, added to the water, stuck to the trails and exposed them as dark bands. The area of trail produced over the prescribed time (measured planimetrically) was used as an index of activity. Table V summarizes the results for worms in filtered lake water only, and in lake water to which had been

TABLE V

The effect of starvation on movement (cm² of mucus trail produced per 3 hours) in freshwater triclads

Species	Treatment†	Mean area‡		Starvation time (days)							Anovar§		
		$\bar{x}$ / Stand. error SE		0	1	3	7	14	21	28	F	d.f.	P
D. lacteum	LW	$\bar{x}$	49·30	30·36	23·54	13·77	12·42	13·68	11·28	19·9	6/63	<0·001
		SE	5·59	3·63	2·16	3·74	1·75	2·15	2·15			
	LWA	$\bar{x}$	55·22	47·36	40·06	38·76	49·58	47·24	38·76	0·8	6/28	>0·05
		SE	11·99	6·99	7·15	5·09	6·47	6·93	5·09			
P. tenuis	LW	$\bar{x}$	18·94	24·67	23·48	41·41	39·46	8·78	8·20	21·3	6/63	<0·01
		SE	4·18	2·09	2·92	4·30	2·73	2·01	1·83			
	LWA	$\bar{x}$	20·14	17·40	19·76	22·14	17·76	8·72	8·40	1·3	6/28	>0·05
		SE	5·89	4·32	5·86	8·27	4·23	3·39	3·13			

† LW = filtered lakewater; $LWA = LW$ + $Asellus$ homogenate. ‡ Based on 10 replicates. § The analysis of variance tests the significance of the trend by considering whether the between-day variance is greater than the between-replicate variance. The analysis was based on the assumption of a completely randomized design, F = variance ratio, df = degrees of freedom, P = probability of rejecting null hypothesis, between-day trend not significant, when it is correct.

added the filtered homogenate of one *Asellus*. *D. lacteum* slowed down during starvation but was activated by the homogenate whereas *P. tenuis* speeded up during starvation (to a max. level at 14 days) and its speed was maintained (up to 14 days) in the presence of the homogenate. These responses are as expected. The homogenate presumably acted as a signal, to both species, of the presence of food. Using a different technique Bellamy and Reynoldson (1974) have obtained similar results to these.

I turn now from behavioural strategies involved in finding food to physiological strategies deployed to conserve body reserves. As an example, Fig. 8 illustrates the respiratory response of the snail, *Planorbis contortus* (Gastropoda; Pulmonata) to starvation. Plotted separately

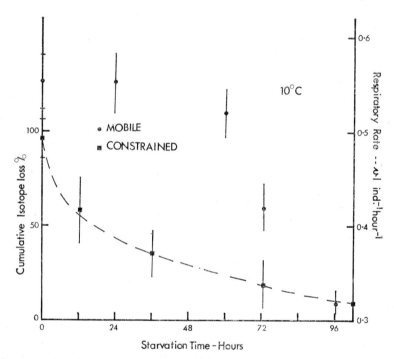

FIG. 8. The effect of starvation on the respiration of constrained and mobile *P. contortus*. Vertical bars give 95% confidence limits and the broken line represents the timecourse of gut emptying (with permission from Calow, 1974a).

are the rates of mobile and constrained snails in cages which prevented movement of individuals and this shows clearly the metabolic expense incurred by searching. The latter falls off as movement reduces in weakened individuals (Calow, 1974a). The cost of movement, however, is offset in *P. contortus* by reductions in pacified (resting) metabolism;

this falling at a progressively reducing rate with starvation time and
following the same time-curve as gut emptying. Hubbell (1971) has
reported a similar response in *Armadillidium vulgare*, a terrestrial
isopod, ascribing part of it to the well-documented phenomenon of
specific dynamic action and another part to the depletion of endogenous
energy reserves. Both these mechanisms are passive manifestations
of depletions in energy supply; a conclusion consonant with the fact
that reductions in pacified respiration follow the same curve as gut
emptying. As in the vertebrates (Garrow, 1974), however, active
control mechanisms may also be involved. Whatever the cause, though,
the effect of this strategy will be the same; reducing energy expenditure
during shortage and off-setting any expense invested in searching.

One process which is often overlooked as a potential element in
metabolic regulation is defaecation. Guts almost always contain food,
some of which will not be digested and absorbed. Consequently the
contents of the alimentary tract may provide a store of potentially
utilizable food which might be exploited during the early stages of
starvation. It is to be expected, therefore, that the rate of passage of
food through the gut should be sensitive to food availability. Figure 9
shows various defaecation strategies used by the snail, *A. fluviatilis*,
in response to food availability. Here the food was labelled with ^{51}Cr,
a γ emitter, which was not appreciably absorbed across the gut wall
(Calow and Fletcher, 1972). Hence the loss rate of isotope from snails
provided a good measure of defaecation rate. When satiated individuals
were transferred from labelled to non-labelled food ^{51}Cr loss, expressed
logarithmically against time, followed a single straight line (*a* in Fig. 9)
with a negative slope. However, in cases where snails were starved
either before or after exposure to ^{51}Cr (lines *b* and *c* in Fig. 9) the single
line broke into two separate linear components. Microscopical exami-
nation of the faeces showed that the second less-steep phase was
associated with the passage of food through the hepatopancreas, an
organ concerned almost exclusively with absorption in this species
(Calow, 1975b). In a continuous feeder synchrony between the rate
of passage of food through each part of the gut is necessary to prevent
blockage but during starvation this becomes less important. Here,
though, a reduced rate of loss of potentially digestible material through
the site of digestion may facilitate the extraction of food from it
(see below).

A similar process has been described recently in woodlice (Hassall
and Jennings, 1975, and pers. comm.). When litter is abundant these
animals feed almost continuously and the gut is usually full. When
litter is less abundant, normally indigestible food is held in a distinct
part of the gut for a special process of digestion effected by cellulases

produced from micro-organisms actually ingested with the food. In this way the passage of food through the woodlouse is slowed down and more nutrient is extracted from the ingesta during times of nutritive stress.

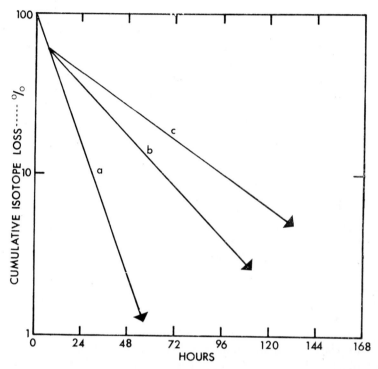

FIG. 9. Defaecation strategies in *A. fluviatilis* as monitored by the loss rate of ^{51}Cr. See text for further explanation. (Data from Calow, 1975b.)

Modification in the pattern of defaecation can thus contribute to fitness. In general, animals must decide whether to hold food in the gut for more efficient exploitation or void it to make room for more easily handled material (see Section XI, B). The latter strategy should be adopted in times of plenty, the former in times of famine.

b. *Response when food is replenished*

It is well known that if food is made available after a period of starvation the ingestion rate rises significantly above normal. However, it is equally well documented that ingestion rate increases at a progressively reducing rate with starvation time ultimately reaching an asymptotic level which reflects the physical capacity of the stomach

(Miner, 1955; Holling, 1966; Brett, 1971; Calow, 1975a; Townsend, 1975). One model accounts for this behaviour by relating appetite solely to the unfilled volume of the gut (Holling, 1966) but evidence like that of Barnett (1953), which showed that rats ate less when sugar was added to their water supply suggests that at least some animals regulate their ingestion rate in terms of the energy content rather than the mass of the food eaten (see also Calow, 1975a). Once again, however, the effect is more important than the cause from a strategic point of view and the effect here is clearly to make good, at least in part, the energy deficit incurred during starvation. Often associated with the increase in ingestion rate, and complementing it, is an increase in the efficiency by which ingested food is digested (Brett, 1971; Calow, 1975a; Windell, 1966). However this phenomenon does not appear to be universal (Conover, 1966).

B. REDUCTIONS IN QUALITY

On the basis of current evidence it now seems that there are no true omnivores in nature, though there may be generalists—eating foods belonging to the same broad class (e.g. animal or vegetable) in direct proportion to their availability. Even this is difficult to substantiate, though, since most of the evidence on which it is based comes from an analysis of stomach contents only, and not food availability. When sufficient information does become available it often exposes strong preferences even in groups where they are least expected. For example, *Ancylus fluviatilis* browses over the algal mat which adheres to submerged surfaces in aquatic habitats. Feeding here involves the scooping action of a radula and might be expected to be indiscriminate. However, a comparison of stomach contents with potential food actually available on the substratum has revealed that diatoms are taken in preference to blue-green algae (Calow, 1973b). Even within the diatom group genera are not eaten in direct relation to their availability (Calow, 1973b).

As yet there is no general consensus on what determines food palatability. All other things being equal (e.g. vitamins, essential aminoacids), however, it seems reasonable that animals should show preferences for energy rich foods. What constitutes an "energy rich food", may of course vary from species to species depending on their ability to locate, capture, ingest and digest the materials. In the extreme case it is of no use to a herbivore that meat is a richer source of energy in terms of calories per gram since, being a herbivore, it is not equipped to deal with meat. For this reason we should not expect to find a direct correlation between the preference sequence of the feeder and the calories/g of the food (c.f. Paine and Vadas, 1969). Alternatively, with foods which are equally easy to locate, capture and ingest we should expect to find a correlation between the "utilizable calories per mouthful of food", an index which expresses how much energy can be absorbed from a unit weight of ingested food, and preference. Figure 10 shows that this is the case for *A. fluviatilis* and I have found similar relationships in *Lymnaea pereger* and *Planorbis contortus*. Indeed food preferences of freshwater snails can be broadly correlated with the ability of these animals to digest different food

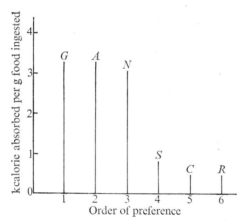

FIG. 10. Nutritive value of food, in terms of potential energy absorbed across the gut wall per weight of food ingested, against preference order of food materials to *A. fluviatilis*. *G* = *Gomphonema*; *A* = *Achnanthes*; *N* = *Navicula*; *S* = *Scenedesmus*; *C* = *Cladophora*; *R* = *Rivularia* (with permission from Calow and Calow, 1975).

types and ultimately with their complement of digestive enzymes (Calow and Calow, 1975).

If, for some reason, the preferred food item becomes temporarily scarce the feeder must switch to another food or starve. Starving animals, therefore, tend to be less discriminating in their diet than satiated ones (Brody, 1945; Calow, 1973b, 1974b). This is also true of animals in an environment where all the food is generally scarce (Emlen, 1973; Ivlev, 1961).

The switching strategy is itself an adaptive device designed to deal with a fluctuating food supply but it may be associated with other mechanisms specifically dealing with the concomitant deterioration in quality; that is, in terms of energy supplied per mouthful. Evidence from the snails, for example, suggests that respiratory rate is reduced in animals fed on poor quality food (unpublished), that less preferred foods are moved through the gut at slower rates thereby increasing the chances of extracting nutrients from them (Calow, 1975b), and that when, in the laboratory, algal food is offered in mono-culture, there is a negative relationship between absorption efficiency and ingestion rate (Calow, 1975a). The latter will tend to maintain a constant flow of energy across the gut wall even in the face of fluctuations in the quality of the food supply and Schindler (1971) recorded the same response in the zooplankton. Detrivores also deploy similar mechanisms as the digestible component of their food becomes diluted with intractable material like lignin. Here dilution usually results in increased consumption rates (Calow, 1975b; Prinslow and Valiela, 1974).

C. CONCLUSIONS

There is abundant evidence both at the whole-organism level and in terms of individual, physiological and behavioural processes that animals can adapt to disturbances in food supply. When this information is viewed in an ecological context it becomes possible to see how and why organisms have become adapted to be physiologically adaptable. As well as there being a physiological pay-off from putting these phenomena in an ecological context there may also be a reciprocal ecological pay-off. The fact that organisms are adaptable in metabolism, for example, undermines the ecological idea that individuals and populations are passive valves in the flow of energy through ecosystems and that they can be modelled with fixed turnover efficiencies (Hubbell, 1971; Calow, 1975a). I shall discuss this more fully in Section XIII.

XII. SOMA *VERSUS* GERM LINE

It should be apparent from the preceding section that the homeostatic regulation of one parameter must, of necessity, lead to alterations in other parameters. For example, the maintenance of a given growth trajectory during disturbance is effected by adjustments in the food gathering and processing activities of animals and in modifications of respiratory output. Food supply disturbance during reproduction leads to a conflict of controls; should reproductive output be preserved at the expense of parental metabolism or *vice-versa*? The nature of this conflict is clearly illustrated in Table VI for the freshwater bug, *Corixa punctata*. Virgin corixids do not lay eggs and though survival is slightly affected by starvation the effect is not marked. Alternatively both starved and fed, fertilized females lay eggs and though there is a significant difference in the egg production rates of these two groups the effect is not marked. However, the survival of starved, egg-laying females is significantly and markedly reduced with respect to their fed counterparts. Thus, reproductive output during disturbance significantly influences parental survival, a feature characteristic of annual but not perennial species (Calow, 1973c). Other evidence showing that the prevention of reproduction significantly increases the longevity of semelparous females is summarized in Table VII.

TABLE VI

The effect of starvation on the egg production and survival of virgin and mated corixids

Type	Treatment†	Eggs/individual/day‡	Longevity§
Virgin	Fed	—	28·1 (± 5·1)
	Starved	—	23·2 (± 4·3)
Mated	Fed	1·66 (± 0·21)	27·3 (± 5·4)
	Starved	1·24 (± 0·18)	12·4 (± 2·2)

† Fed subjects were given superabundance of spinach; starved subjects were kept in filtered lakewater with no food.
‡ Averages, measured from the initiation of oviposition to death, with 95% confidence limits. The results are significantly different; $t = 4·20$ for 14 $d.f.$, $P < 0·05$.
§ Average time (days) from initiation of oviposition to death with 95% confidence limits. Results for the two groups of virgins are not significantly different ($t = 0·57$, for 14 $d.f.$ $P > 0·05$). Results for the two groups of mated females are significantly different ($t = 18·86$ for 14 $d.f.$ $P < 0·001$).
General note. Fifteen replicate individuals were used in each treatment. Subjects were collected from the field in November, before fertilization. Ovarian development was initiated at this time by culturing females at 10°C on a 16 hour, photoperiod (see also Young, 1965). Plastic weed was used as an oviposition site.

The physiological basis of the relationship between fecundity and longevity will be discussed in more detail below. For the present my main concern will be with the evolutionary issue. The problem is whether to breed "come what may" and in so doing put the parent at risk or to build into the system mechanisms, similar to those artificially imposed on the virgin corixids and apparently a natural feature of iteroparous (repeated breeding) species, which "switch reproduction off" during stress and so makes reproductive output subservient to parental well-being. The solution must depend on the relative merits of allowing the genetic message to be carried through a difficult period by an old "experienced" or young "virile" stock. This is likely to depend on the biology of the organism and the nature of the environment in which it finds itself (Cole, 1954). It might be predicted for example in "r" situations of selection, where disturbance is unpredictable, that the emphasis should be on survival by sheer weight of

numbers; reproduce come what may. Alternatively, in "K" situations
of selection, when competition is keen and resources are limiting,
"experience" will count for more than fecundity so that here one
would expect the evolution of rigid controls on reproductive output.
Certainly, iteroparity seems to be more a feature of mature communities
(Emlen, 1973).

TABLE VII

*Evidence showing that the prevention of reproduction significantly increases the
longevity of semelparous females (modified from Calow, 1973c)*

	Evidence	Source
1.	Minnows normally die after spawning, but non-spawners may live an extra year	Markus (1934)
2.	Gonadectomised salmon do not spawn and are able to return to the sea	Calow (1973c)
3.	Heat sterilised, unmated and ovariless *Drosophila* live significantly longer than mated females	Maynard Smith (1959)
4.	Virgin female bugs (Hemiptera) and beetles (Coleoptera) live longer than mated females	Clarke and Sardesai (1959); Murdoch (1966)
5.	Observations 3 and 4 are typical for most insects	Clark and Rockstein (1964)
6.	Death in some annual plants (e.g. tomatoes) can be postponed indefinitely if reproduction is prevented	Heath (1957)

XIII. MODELS OF METABOLISM

In the foregoing it has been possible to distinguish between two
senses of the word "adaptation"; that referring to a certain level of
operation in the metabolic parameters and that referring to the regulatory phenomena elicited in individual organisms after disturbances.
For clarity I have referred to the latter as "adaptability". As I have
tried to show, both processes depend ultimately on the modifying
influence of the environment, but in each case this occurs in slightly
different ways. Levels of adaptation, for example, depend on environ-
mentally mediated shifts in the genetic make-up of a population through
the differential replication of genes. Alternatively that aspect of the
adaptive strategy which is deployed under a particular set of environmental conditions depends upon modification in the expression of the
genome in the individual organism. In the first place environmental
stimuli act as "copy editors" scoring out badly written messages and

underlining well written ones; in the second place they act more like "producers" bringing into focus this or that part of the genetic picture.

The way organisms respond to *proximate* environmental stimuli must *ultimately* be explicable in the way the environment has moulded the level of adaptation in ancestral forms. There are in fact several ways in which proximate stimuli may influence organisms; this reflecting differences in design and differences in the way systems have come to be designed. First, environmental disturbances may involve an *active* regulatory response similar to that seen in automatically controlled missiles which home in to a target despite evasive tactics and thermostatically controlled baths which maintain a constant internal temperature despite fluctuations in external temperature. Here the desired state is explicity programmed into the system as a "set point" and sensors continuously measure how well the mechanism is doing what is required of it. Secondly, disturbances may involve a *passive* regulatory response which is written into the design but only as an implicit consequence of the organization of the system; something like the regulatory response elicited after deflecting a swinging pendulum or a spinning top. There are no explicit programmes here and yet it is implicit to the way the pendulum and top work that they should tend to resist disturbance. Finally, environmental stimuli may bring about non-regulatory responses; the sort of thing that would happen by "jamming" the "homing device" of the missile, raising the external temperature of the bath to such a level that the thermostat could never cope or by cutting the string of the pendulum. In developing a theory of metabolism, therefore, we must be able to take these different forms of response into account. Ultimately this means that we must be able to build models of metabolism which respond to environmental inputs in the appropriate ways.

The most straightforward model of metabolism simply considers the organism to be an energy store (G) into which flows organized energy (A) and out of which flows disorganized heat energy (H). Here changes in biomass (dG/dt) depend on the difference between A and H:

$$dG/dt = A - H \qquad (5)$$

Equation 5 could be made more realistic by putting the input as ingested rather than absorbed energy and by including faeces (F), excreta (U) and secreta (S) in the output term. It would also be necessary to include reproduction (Rep) as another component of energy loss because as far as the individual organism is concerned gamete loss is as real an energy drain as excretion and secretion. Equation 5 thus becomes:

$$dG/dt = I - (F + H + Ex + S + Rep) \qquad (6)$$

In this model the level of adaptation is fixed by the parameter values assigned to the output terms and to the value associated with I under superabundant feeding conditions. For this most simple situation environmental disturbances can only operate directly through I, though indirectly, disturbances may influence the other parameters, e.g. as temperature, humidity and salinity effects. The resulting response of the system to any of these disturbances would be of a non-regulatory type. For example, because the output terms are fixed, changes in I would result in proportional changes in dG/dt and this means that there should be a stoichiometric relationship between input and useful output irrespective of the level of the input. Evidence from Section XI shows that this is clearly an over-simplified view of metabolism.

Considering the model in general input-output terms we can make it more realistic by putting the output parameters as some function of body size (G):

$$H = f(G) \qquad (7)$$

Substituting in Eqn. 5 we obtain:

$$dG/dt = A - f(G) \qquad (8)$$

and the model now contains an implicit regulatory mechanism. This is best illustrated by rewriting the integral of Eqn. 8 in block diagram form (Fig. 11). Here the circle with a sum sign in it is a comparator

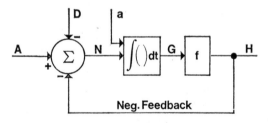

Fig. 11. Block diagram of Eqn. 8.

which sums $-H$ and A to give the net energy input N. Integrating, or summing N with respect to time ($\int(\ )dt$) and including the initial size at birth (a) gives G (organismic size) and this multiplied by f (the respiratory parameter) gives the output term H. The latter feeds back to the comparator and the whole process continues iteratively throughout life. Environmental disturbances (D) subtract from growth and are modelled as a generalized negative input signal to the comparator.

Notice that the model depicted in Fig. 11 contains a crucial device for automatic regulation; the negative feedback loop. In fact there is no physical mechanism in the system which actually corresponds to this loop; rather it is due to the cancelling out of forces within the system and for this reason it is often referred to as *fictitious feedback*. This sort of behaviour characterizes any open system (physical, chemical or biological) in which there is a store interposed between input and output and in which output is functionally related to the size of this store. Because of the fictitious feedback, Eqn. 8 models limited passive regulation in G and dG/dt during disturbance and this corresponds to the pendulum paradigm. Reductions in A, for example, cause reductions in G which lead to lowered output, H, and ultimately to a rise in N. G and dG/dt are therefore buffered to a limited extent against fluctuations in energy supply.

Though Fig. 11 is a very simple model of metabolism it does, nevertheless, form the basis of both the familiar energy budget equation

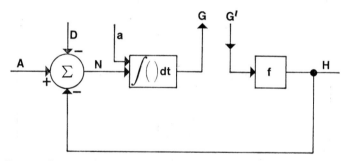

FIG. 12. Frequently used modification of the model in Fig. 11.

and the von Bertalanffy growth equation (Bertalanffy, 1960). These usually make A a size-dependent term and put both the inputs and outputs as non-linear functions, allometrically related to body size ($f(G^x)$, where $x \approx 0{\cdot}5$ to $1{\cdot}0$). Even so, the resulting models do retain the major features of Eqn. 8 and particularly the passive regulatory response. Frequently, however, especially in ecology, the feedback channel is broken to admit external data. More often than not G or a related parameter is taken out of the model and G', obtained from observation, is put in to operate the respiratory sub-system instead (see Fig. 12). Failing to use the system as an autonomous iterative device has two effects. First, the intrinsic control properties of the system are destroyed and second, since G is often compared with G' as a means of checking the accuracy of the model, a degree of circularity is introduced into the argument. Figure 12 clearly shows that G and G' must always be related. Though these moves may be justified in using the model to describe particular individuals, in a particular place and at a particular time they seriously undermine the generality of the approach.

Passive control, through fictitious feedback, can clearly account for some of the ways growth and size are buffered against external disturbances but it is not possible to account for all metabolic adaptability in this way. Passive control is not very precise, for example, and though it can minimize change after disturbance it does not account for continued control during perturbation and for the tracking phenomena already described. If the growth trajectory is such an important component of fitness (Section V) then it is likely that this parameter is controlled more precisely. At a physiological level there certainly does seem to be precise active control over feeding, if not respiration

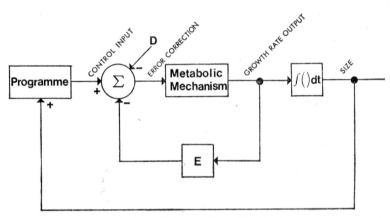

Fig. 13. A cybernetic model of metabolism with real feedback.

(Section XI) and there is evidence for real feedback paths involving special regulatory molecules and for a definite growth programme (Weiss and Kavanau, 1957; Needham, 1964). This suggests a shift from the "pendulum" to the "thermostat" paradigm.

The basic model is illustrated in Fig. 13. This differs from passively controlled open systems by having an explicit programme (sometimes known as a motivator), a physically definable comparator and a real feedback path. Notice that the input of this system is in terms of information, the command from the controlling programme (in this case a "desired growth rate" signal) and not energy or matter. The output is the response of the metabolic mechanism in terms of energy inflow and outflow over a given time and is in terms of actual growth rate. This feeds back to the comparator when it is subtracted from the command input. The difference passes out to the metabolic mechanism and being proportional to the original misalignment attempts to compensate for it. Interposed between the output and the comparator is a feedback parameter, E. This indicates that some real feedback circuitry is involved which, being real, operates at a certain level of efficiency. When $E = 1$, for example, the feedback mechanism is 100% efficient in transferring information but when $E < 1$ information is lost during feedback. As well as feeding back to the comparator, the output, after integration, feeds back positively to the motivator. This is because growth is a size dependent process which must be able to take into account the present state of the system in terms of size. Notice, as in the last model, environmental disturbances are modelled generally as a negative input signal to the comparator.

Once again Fig. 13 only provides an extremely simplified model of metabolism. As a general basis it has nevertheless been exploited by Hubbell (1971) to model the bioenergetics of invertebrates to various degrees of realism. From it there are two possible routes of research; one quantitative, the other qualitative. The quantitative route involves using the model, in simulation, for predictive and descriptive purposes, and requires that specific parameter values be associated with each sub-system. Here it has to be admitted that it is not always easy to identify and measure the input generator and feedback parameter in biological systems and often each and all of the parameters may vary continuously with time. Hubbell has made some suggestions on how the nature of the input might be defined in living organisms (Hubbell, 1971).

The qualitative route of research from Fig. 13 involves using the principles and practices of systems theory (Calow, 1976) to work out the input-output relationships of the system and the possible effect of environmental disturbance by reference to its organizational logic only. Thus, it is possible to use the rigorous rules of operational calculus to make explicit the consequences, in terms of the system as a whole, of connecting particular sorts of sub-systems in particular sorts of ways. It is also feasible, though not so straightforward, to start with input-output information and then to progress to the organizational circuitry of the system. The former is known as the synthetic approach, the latter as the analytic. Often it is possible to use both methods on the same problem.

To illustrate the "systems approach" consider Fig. 14. This shows a model of metabolism, modified from Hubbell (1971), which incorporates both the passive size dependency of the metabolic parameters and active feedback. The notation is as usual. M is a motivator which generates a "desired growth rate input" (DGR). This is compared with actual growth rate (AGR) at comparator 1 and on the basis of the difference an error-correcting signal feeds back negatively on H

(respiration reduces with starvation, Section XI, A, a) and positively on A (the tendency to feed increases with starvation, Section XI, A, b). Each of these parameters is size (CS) dependent. *Rep* represents the reproductive system which has a negative effect on energy inflow at

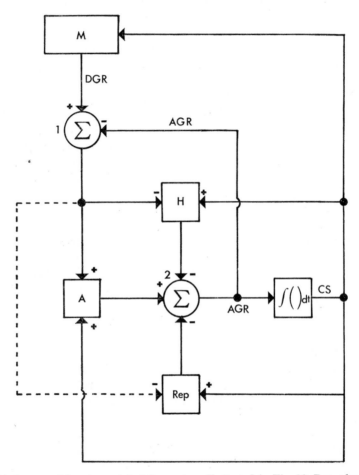

FIG. 14. A more elaborate model of the system illustrated in Fig. 13. Reproduction is included here as an element of catabolism. The broken line represents a feedback path which may or may not be operational. (Modified from Calow, 1973c.)

comparator 2. It is in turn influenced positively by size (since reproductive output generally increases with size; Section V), and negatively by a signal proportional to metabolic well-being. However, this feedback path is broken to allow for the two possibilities discussed in the last section; in some species reproduction continues irrespective of parental well-being (feedback broken) whereas in others it is rigidly controlled by the parental system (feedback operational).

Having constructed the model, the next stage is to investigate its properties by looking at how its output is influenced by inputs of

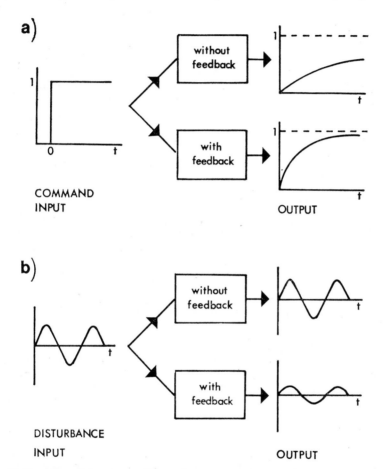

Fig. 15. Response of the system illustrated in Fig. 14 to a standard step input in the command variable (a) and a standard sinusoidal disturbance (b). These signals are commonly used for investigating the properties of linear systems.

various sorts. In theory any input signal will do but in practice a few inputs of standard form are used for convenience (Calow, 1976). Mathematical functions representing these generalized "test inputs" are applied to the equations representing the system and the output function is obtained. There is nothing magical about these procedures. They simply amount to making clear, by mathematical means, certain consequences of the design features of systems which might otherwise remain hidden in the intricacies of the model.

For example, with the models in Fig. 14 it has been possible to show that when generalized, control signals are put into each system the output responds more accurately and more promptly if the feedback circuit is intact (Fig. 15a). Also, when a generalized disturbance signal is put into each system the one with feedback damps out disturbance more effectively than the one without feedback (Fig. 15b). Since the extent to which disturbance impinges on metabolism is likely to be reflected in survival, reproduction without parental controls will put the parent at risk. Reproduction without feedback, therefore, increases the chances of parental death and as such can be considered as an element of senescence. These predictions are fully consistent with the results presented in the last section and begin to focus on the possible physiological mechanisms behind them. The evolution of control paths from the parental to the reproductive system are clearly important in the development of an extended life-history and iteroparity.

Models of the type illustrated in Fig. 14 incorporate a number of important features. First, they take into account the fundamental aspects of metabolism outlined at the beginning of this section. Second, they offer a general framework which can be elaborated to improve *realism* on the one hand and *precision* on the other. Finally, by setting the physiological strategies within the cybernetic framework of a programmed and programmable system they look outwards to the problem posed by the immediate environment and backwards to evolutionary experience. It is in this context, then, that the physiologist can take full cognizance of ecology and evolution. A similar approach using more complex models involving "optimal control theory" has been followed by León (1976).

XIV. Postscript on Ageing

In an article on the development and mode of operation of "metabolic machines" it is not possible to ignore the problem of wear and tear, fall-off in performance, and ultimate breakdown. This manifests itself as a progressive deterioration in homeostatic ability and as an increased likelihood of dying. We refer to the sum total of these effects in organisms as the process of "ageing" or of "senescence". From the point of view of a mechanistic view of life it is important to consider whether an "ageing programme" is written explicitly into the design of living things or whether senescence is simply due to the random accumulation of damage within the system. This question, as with the others that have been considered in this article, has a physiological, an evolutionary and an ecological component. I put it in a postscript since it is indisputable that death usually occurs well before the onset of senescence in natural ecosystems (e.g. by predation and disease) so that ageing is not usually of direct concern for the ecologist.

From the point of view of the "accident *versus* design" aspects of senescence there are three possible theories, not two as suggested above. Ageing may represent: (*a*) the random accumulation of accidental damage; (*b*) the acting out of a genetically specified series of events; (*c*) genetically controlled accumulation of accidental damage. Theory *a* implies a purely fortuitous process. Theory *b* suggests that specific "suicide instructions" are written into the genome and that these are switched into operation at a specific time in the life of an organism. Theory *c* is intermediate between both *a* and *b* implying that organisms are made susceptible to the accumulation of certain sorts of accidental damage at a certain rate because of the way they are designed. Of course, theories *a* and *c* are usually difficult to distinguish in that any design can always be said to be particularly susceptible to certain sorts of damage. In fact this difference hinges not on the outward manifestation of the way organisms work but on the intentions of the designer. The difference between the natural rusting of metal objects (just because they are metal) and the rusting which manufacturers of metal objects might "engineer" to keep up the demand for their product illustrates the sort of difference envisaged between theories *a* and *c*. It is implicit to the latter strategy that "designed decay" is indistinguishable *as a process* from "natural decay" and *vice-versa*. Ageing, therefore, can always be seen as a more or less direct consequence of organismic design irrespective of which of the three theories turn out to be correct. The remaining question, however, is whether ageing is an "intentional" (theories *b* and *c*) or "unintentional" (theory *a*) consequence of organismic design, and this is the evolutionary issue. Because it is possible for decay to be design-like without actually being designed it is not permissible to argue from the design-like behaviour of senescence (e.g. the within-species constancy of senescent changes and their timing) to the idea that there is any *positive* selection for it (e.g. the argument by Burnet, 1974). This question must be decided on the basis of evolutionary logic and possibly selection experiments.

Are there any ecological circumstances in which there might be a positive selection pressure for senescence and death? In terms of sheer evolutionary logic anything which reduces the total potential fecundity of an individual, as decay and death must do, will have a negative not a positive effect on fitness; life is about living not dying. Alternatively, it is often argued that death is needed in the evolution of life to make room for new adaptations and also to prevent less viable and fecund parents coming into competition with more viable and fecund offspring. In the first place, though, it should be remembered that there can be death without senescence and, as discussed above, "ecological" death is more common than "physiological" death in nature. In the second place it is difficult to see how selection might operate on the basis of interactions between generations. It is certainly true that in some circumstances and for many organisms death of the parent might benefit the offspring by reducing the intensity of intraspecific competition. But the problem here is that the gene for dying would not only advantage the progeny carrying it but also other conspecific progeny not carrying it. That is, there would be a general alleviation of competition through the programmed death of an adult not necessarily specifically restricted to the genetic lineage of which the character was part. Thus, apart from the special circumstances of kin selection, where there is strong "family" grouping (as in social insects), the process would not be able to select out the gene for dying in the usual way. Consequently, without invoking group selection (Wynne-Edwards, 1962) the only alternatives are two variants of theory *b*. That is, either to view ageing as an inevitable consequence of any biological organization or as a side effect of specific design features which at one time in the life-cycle have a sufficiently positive effect on fitness to outweigh negative effects expressed later on. Since certain biological systems do apparently escape ageing, e.g. germ cells, asexual lines, "vegetative cuttings" etc., the most reasonable alternative would seem to be the second and several specific evolutionary theories of senescence are based on this conclusion (Medawar, 1952; Williams, 1957; Hamilton, 1966; Edney and Gill, 1968). Though these differ in detail they are based on the same basic logic and the essential points common to them all are as follows: The older an organism gets the more chance it has of dying by "artificial means", predation, disease etc. There comes a point in the life-cycle, therefore, beyond which the probability of living on becomes extremely low unless, of course, organisms are given special care as in laboratories or the "Welfare State". Because of this, genes that express themselves late in life escape partly or wholly from the force of selection. It is demonstrable, therefore, that any genes or design features which confer a reproductive advantage early in life will be selected for even if they have ill-effects later.

According to most of these theories senescent changes ought to be a mixed bag of side-effects from *any* genes that happen to have favourable effects in younger stages. On the contrary senescent changes seem to follow such a well-defined course both within and between species (Comfort, 1964) that this can hardly be the case. Is it possible, therefore.

that senescence is modulated by a single design feature which is of uttermost and general importance in the evolutionary competent part of the life-cycle of most living beings but which has a fairly constant and predictable effect in the "selection shadow"? One possibility stems from the control of growth and organismic size. This is likely to be under strong selection to slow down and stop in favour of reproduction at the appropriate part of the life-cycle (Section V) and there are also reasons for expecting growth to be strictly controlled at the organ and tissue level (Cairns, 1975). Ultimately, at all levels, growth control depends on the regulation of cell turnover (Bullough, 1967) and the build up of non-dividing cells. These post-mitotic cells are presumably especially vulnerable to the probability of damage by use and by somatic mutation in its widest possible sense (Curtis, 1966; Orgel, 1973). Thus, as the level of post-mitotic cells increases within the body the probability of fatality through the accumulation of damage also increases. It is this side-effect of growth control which will become particularly obvious in the "selection shadow", which can therefore be assimilated into the genetic system and which could form the fundamental basis of senescence in most living systems. If this is the case old ideas concerning the reciprocal but continuous relationship between growth and ageing (Bidder, 1932), the possible reversibility of senescence by "degrowth" and regeneration (Child, 1915) and the notion of constant "growth energy", which, once spent, results in death (Rubner, 1908), take on new meaning.

Using the general framework of a theory of senescence based on growth control it is now possible to bring into account the particular theory of "reproductive senescence" discussed in Sections XII and XIII. On the hypothesis that "reproductive senescence" operates in conjunction with environmental disturbance to accelerate the normal process of senescence rather than acting as an independent mechanism, it is necessary to identify means whereby this could occur. There seem to be at least two possibilities, each compatible with the model illustrated in Fig. 14. First, under conditions of nutritive stress the reproductive system might use up tissue which being in a post-mitotic condition would be irreplaceable. This would inevitably weaken the system and thereby make it even more sensitive to general, environmental stress. Figure 16, for example, shows how the lipid (storage compartment), flight muscles (functional compartment) and gonads (reproductive compartment) of mated and virgin corixids respond to starvation (see also Table VIII). In the mated individuals gonadial homeostasis is maintained at the expense of both the storage and the functional compartments. Most insect tissues are post-mitotic (Bullough, 1967) so it is likely that material lost from the functional compartment would

not be replaced. Hence the parent system becomes irreversibly weakened. In the virgins, homeostasis is maintained at the expense of
the gonadial and storage compartments. Hence weakening here is not

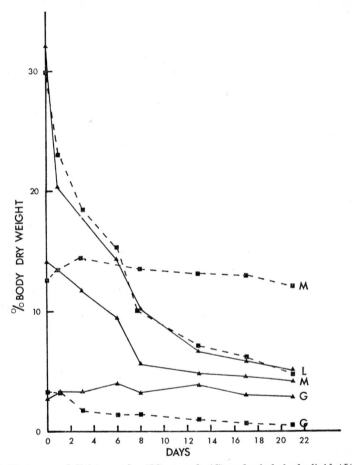

FIG. 16. Response of flight muscles (*M*), gonads (*G*) and whole-body lipid (*L*) to increasing starvation (days). The dry weight of each of these components is expressed as a
% of the whole-body dry weight. Squares linked by broken lines are the average results
of the virgins ($N = 8-10$), triangles linked by solid lines are the results for the mated
females ($N = 4-8$). For clarity confidence limits are not shown but standard errors
ranged between $\pm 4\cdot3-12\cdot7\%$ of the mean.

so pronounced as when the reproductive apparatus is functional and
this difference could account for the differences in longevities noted
between mated and virgin bugs in Table VI.

The other mechanism whereby reproduction is able to interact with
senescence is less direct and more difficult to substantiate. Here it is

envisaged that reproduction adds more stress to an already stressed organism, and thereby, accelerates the rate at which damage accumulates within the parental system. It is well known that metabolic stress can reduce life-span and evidence from autopsy shows that animals treated in this way do not die from a single cause, but from the whole gamut of causes usually associated with senescence (Curtis, 1966). That reproduction does not cause death *per se* but interacts with environmental disturbance is demonstrated by the fact that most annual organisms can be kept alive, whilst breeding, for periods longer than a year under good laboratory conditions (Calow, 1973c). For example,

TABLE VIII

Regression equations relating the relative weights of various body compartments of C. punctata *to starvation time (days). A regression coefficient not differing significantly from zero indicates good homeostasis*

	Log_{10} (compartment wt./body wt.) $= a+b$ (starvation time)	N	b sig. diff. from zero $P \leqslant 0.05$
Mated			
Gonads	$y = 0.5422 + 0.0011\ (\pm 0.0134)\ x$	61	
Flight muscle	$y = 1.1213 - 0.0271\ (\pm 0.0006)\ x$	77	*
Lipid	$y = 1.4221 - 0.0385\ (\pm 0.0112)\ x$	63	*
Virgin			
Gonads	$y = 0.4693 - 0.0367\ (\pm 0.0113)\ x$	64	*
Flight muscle	$y = 1.1357 - 0.0017\ (\pm 0.0035)\ x$	51	
Lipid	$y = 1.4046 - 0.0398\ (\pm 0.0118)\ x$	60	*

* = significant difference; b = regression coefficient; a = constant.

egg-laying *D. lacteum* (usually annual in nature; Reynoldson, 1966) lives approximately 200 days when fed bi-weekly under laboratory conditions as opposed to 500 days or more when fed daily. However, heads surgically removed from both the stressed and non-stressed egg-laying adults, and thereby separated from the gonads, do not differ in longevity, living on after transection for about 100 days. Since these explants do not feed but do regenerate a tail, then depletion of reserves cannot be a significant component of the deleterious effect of reproduction. Alternatively, the tails of stressed and non-stressed adults, which in this species are not able to regenerate, do appear to differ significantly in longevity, living on for approximately 100 days when they are from the non-stressed adult and dying after a short time when they are from the stressed adult. In both cases eggs continue to be produced for some time after transection. Therefore at least

C

three factors are involved in reproductive senescence: environmental stress, reproductive activity and reduced cell turnover. I suggest that the only way to explain these data is to postulate an interaction between reproduction and environmental stress which damages the parental system. However, this damage only expresses itself as senescence and death in non-dividing tissue. "Rejuvenescence" is therefore accomplished either by "flushing out" or "dilution" of damage facilitated by the increased cell turnover needed for regeneration.

These examples substantiate the theoretical framework developed in Section XIII. Again, the conclusion is that the physiological mechanisms controlling reproductive output, particularly under stress, will be crucial in the evolution of phenology and longevity. Since this sort of control is unlikely to be all-or-none but graded and since as a matter of theory (Section XIII) the influence of reproduction on longevity will be graded in the same way, what is needed is some sort of easily measured index of the potency of this control. Then, it may be possible to add yet another physiological parameter to the ecology of life-cycles.

To summarize: Life-span is not selected for intentionally. What is under strong selection pressure is when to reproduce and whether reproduction should be made sensitive to parental well-being or not. These objectives are met by selection on the whole gamut of physiological mechanism which are involved in the conversion of input energy to G (Fig. 2) and in the interaction between G and Rep. The ageing and senescence pheonomena develop as a side-effect of selection on these characters. Ageing and senescence, therefore, take on a design-like appearance without actually having been subjected to explicit selection. Guthrie (1969), arguing for a theory of senescence which involved explicit selection, pointed out that the need for early reproduction in some ecological circumstances (e.g. in certain habitats) may provide the required positive selection pressure. I suggest that this theory, though superficially correct, puts the cart before the horse. The fact that there may be positive selection for early reproduction does not in itself explain why this should involve accelerated senescence. Guthrie mistakes the correlation of these two characters with their equivalence. I suggest that early reproduction is usually associated with accelerated senescence because the physiological adaptations necessary for the former almost invariably bring about the latter.

XV. Conclusion

There is no one metabolic recipe for survival. The essential function of metabolism is that it catalyses the transmission of genetic information and it is only in this sense that it can be judged efficient. Though there is a general relationship between energy flowing into reproducta and fitness this need not be the case. Under the appropriate ecological conditions the production of a small number of large gametes may be a more successful strategy than the production of a larger number of small gametes. Some organisms grow big, others grow quickly, others have a high growth efficiency and yet others command a high degree of metabolic homeostasis. There is a surprising diversity of physiological strategies and it is only by considering these in the context of where "metabolic machines" work that they can be fully understood in terms of selection.

ACKNOWLEDGEMENTS

The experimental work on planaria, described in Sections XI and XIV, was supported by a Natural Environmental Research Council grant, no. GR3/2318.

REFERENCES

Adolph, E. F. (1951). Some differences in responses to low temperatures between warm-blooded and cold-blooded vertebrates. *Am. J. Physiol.* **166**, 92–103.

Alexander, R. McN. (1975). Integrated design. *Am. Zool.* **15**, 419–425.

Anderson, J. F. (1970). Metabolic rates of spiders. *Comp. Biochem. Physiol.* **33**, 51–72.

Anderson, J. F. (1974). Responses to starvation in the spiders *Lycosa lenta* (Hentz) and *Filistata hibernalis* (Hentz). *Ecology* **55**, 576–585.

Barnes, H. (1962). So-called anecdysis in *Balanus balanoides* and the effect of breeding upon the growth of the calcareous shells of some common barnacles. *Limnol. Oceanogr.* **7**, 462–473.

Barnett, S. A. (1953). Problems of food selection by rats. *Anim. Behav.* **1**, 159.

Bellamy, L. S. and Reynoldson, T. B. (1974). Behaviour in competition for food amongst Lake-dwelling triclads. *Oikos* **25**, 356–364.

Berg, K. and Ockelman, K. W. (1959). The respiration of freshwater snails. *J. exp. Biol.* **36**, 690–708.

Bertalanffy, L. von. (1960). Principles and theory of growth. *In* "Fundamental Aspects of Normal and Malignant Growth" (Ed. W. W. Nowinski). Elsevier, Holland, pp. 137–259.

Bertalanffy, L. von. (1968). "General Systems Theory." Braziller, New York, 289 pp.

Bidder, G. P. (1932). Senescence. *Br. Med. J.* **2**, 583–585.

Bonner, J. T. (1974). "On Development". Harvard University Press, Cambridge, Massachusetts, 282 pp.

Boucot, A. J. (1976). Rates of size increase and of phyletic evolution. *Nature (Lond.).* **261**, 694–695.

Brett, J. R. (1971). Satiation time, appetite and maximum food intake in Sockeye Salmon (*Oncorhynchus nerka*). *J. Fish. Res. Bd. Can.* **28**, 409–415.

Brodie, P. F. (1975). Cetacean energetics, an overview of intraspecific size variation. *Ecology* **56**, 152–161.

Brody, S. (1945). "Bioenergetics and Growth." Reinhold Publishing Corp., New York, London.

Brooks, J. L. and Dodson, S. I. (1965). Predation, body size and composition of plankton. *Science* **150**, 28–55.

Bullough, W. S. (1967). "The Evolution of Differentiation." Academic Press, London and New York, 206 pp.

Burnet, F. M. (1974). "Intrinsic Mutagenesis. A Genetic Approach to ageing." MTP, Lancaster. 244 pp.

Cairns, J. (1975). Mutation selection and the natural history of cancer. *Nature (Lond.)* **225**, 197–200.

Calow, P. (1973a). On the regulatory nature of individual growth: some observations from freshwater snails. *J. Zool. Lond.* **170**, 415–428.

Calow, P. (1973b). The food of *Ancylus fluviatilis* Müll., a littoral stone-dwelling herbivore. *Oecologia* **13**, 113–133.

Calow, P. (1973c). The relationship between phenology, fecundity and longevity: a systems approach. *Am. Nat.* **107**, 559–574.

Calow, P. (1974a). Some observations on locomotory strategies and their metabolic effects in two species of freshwater gastropods, *Ancylus fluviatilis* Müll. and *Planorbis contortus* Linn. *Oecologia* **16**, 149–161.

Calow, P. (1974b). Evidence for bacterial feeding in *Planorbis contortus* Linn. (Gastropoda: Pulmonata). *Proc. malac. Soc. Lond.* **41**, 145–156.

Calow, P. (1975a). The feeding strategies of two freshwater gastropods, *Ancylus fluviatilis* Müll and *Planorbis contortus* Linn. (Pulmonata) in terms of ingestion rates and adsorption efficiencies. *Oecologia* **20**, 33–49.

Calow, P. (1975b). Defaecation strategies of two freshwater gastropods, *Ancylus fluviatilis* Müll. and *Planorbis contortus* Linn., with a comparison of field and laboratory estimates of food absorption rate. *Oecologia* **20**, 51–63.

Calow, P. (1976). "Biological Machines". Arnold, London.

Calow, P. and Calow, L. J. (1975). Cellulase activity and niche separation in freshwater gastropods. *Nature (Lond.)* **255**, 478–480.

Calow, P. and Fletcher, C. R. (1972). A new radiotracer technique involving 14C and 51Cr for estimating the assimilation efficiency of aquatic primary consumers. *Oecologia* **9**, 155–170.

Calow, P. and Jennings, J. B. (1974). Calorific values in the phylum Platyhelminthes: the relationship between potential energy, mode of life and the evolution of entoparasitism. *Biol. Bull.* **147**, 81–94.

Cathcart, E. P. (1953). The early development of the science of nutrition. *In* "Biochemistry and Physiology of Nutrition", vol. 1 (Eds. G. F. Bairne and G. W. Kidder). Academic Press, New York and London, pp. 1–16.

Child, C. M. (1915). "Senescence and Rejuvenescence". Chicago University Press, Chicago, 480 pp.

Clark, A. M. and Rockstein, M. (1964). *In* "Physiology of Insecta", vol. 1 (Ed. M. Rockstein). Academic Press, London and New York, pp. 227–281.

Clarke, K. U. and Sardesai, J. B. (1959). An analysis of the effects of temperature on growth and reproduction of *Dysdercus fasciatus* Sign (Hemiptera, Pyrrhocoridae). *Bull. ent. Res.* **50**, 387–405.

Clarke, M. F. and Smith, A. (1938). Recovery following suppression of growth in the rat. *J. Nutr.* **15**, 245–256.

Cole, L. (1954). The population consequences of life history phenomena. *Q. Rev. Biol.* **29**, 103–137.

Coleman, W. (1971). "Biology in the Nineteenth Century. Problems of Form, Function and Transformation". John Wiley and Son Inc., New York, London, 187 pp.

Comfort, A. (1964). "Ageing, The Biology of Senescence". Holt, Rinehart and Winston, New York, 365 pp.

Conover, R. J. (1966). Factors affecting the assimilation of organic matter by zooplankton and the question of superfluous feeding. *Limnol. Oceanogr.* **11**, 346–354.

Cummins, K. W. and Wuychek, J. C. (1971) Caloric equivalents for investigations in ecological energetics. *Mitt. Int. Ver. Theor. Angew. Limnol.* **18**, 1–158.

Curtis, H. J. (1966). "Biological Mechanisms of Ageing". Charles C. Thomas, Springfield, Illinois, 133 pp.

Derickson, K. W. (1976). Ecological and physiological aspects of reproductive strategies in two lizards. *Ecology*, **57**, 445–458.

Edney, E. B. and Gill, R. W. (1968). Evolution of senescence and specific longevity. *Nature (Lond.)* **220**, 281–282.

Emlen, J. M. (1973). "Ecology: An Evolutionary Approach". Addison-Wesley Publishing Co., Massachusetts, California, London, Ontario, 493 pp.

Engelman, M. D. (1966). Energetics, terrestrial field studies and animal productivity. *In* "Advances in Ecological Research", vol. 3 (Ed. J. B. Cragg). Academic Press, New York, pp. 73–115.

Etienne, A. S. (1972). The behaviour of *Aeschna*. *Anim. Behav.* **20**, 724–731.

Galbraith, M. G. (1967). Size selective predation on Daphnia by rainbow trout and yellow perch. *Trans. Amer. Fish. Soc.* **96**, 1–10.

Garrow, J. S. (1974). "Energy Balance and Obesity in Man". North Holland Publ., Amsterdam, London, 335 pp.

Gerking, S. D. (1959). Physiological changes accompanying ageing in fishes. *In* "CIBA Foundation Colloquia on Ageing", vol. 5 (Eds. G. E. W. Wolstenholme and M. O'Connor). J. & A. Churchill Ltd., London, pp 181–211.

Golley, F. B. (1968). Secondary productivity in terrestrial communities. *Am. Zool.* **8**, 53–59.

Guthrie, R. D. (1969). Senescence as an adaptive trait. *Perspect. Biol. Med.* **12**, 313–324.

Hairston, N. G., Smith, F. E. and Slobodkin, L. B. (1960) Community structure, population control, and competition. *Am. Nat.* **94**, 421–425.

Haldane, J. B. S. (1928). On being the right size. *In* "Possible Worlds". Harper, New York, pp, 20–28.

Hall, D. J., Cooper, W. E. and Werner, E. E. (1971). An experimental approach to the production dynamics and structure of freshwater animal communities. *Limnol. Oceanogr.* **15**, 839–928.

Hallam, A. (1975), Evolutionary size increase and longevity in Jurassic bivalves and ammonites. *Nature (Lond.)* **258**, 493–496.

Hamilton, W. D. (1966) The moulding of senescence by natural selection. *J. theor. Biol.* **12**, 12–45.

Hassall, M. and Jennings, J. B. (1975). Adaptive features of gut structure and digestive physiology in the terrestrial isopod *Philoscia muscorum* (Scopoli) 1763. *Biol. Bull.* **149**, 348–364.

Heath, O. V. S. (1957). Ageing in higher plants. *In* "The Biology of Ageing", Symposium of the Institute of Biology No. 6 (Eds. W. B. Yapp and G. H. Bourne). Church Army Press, Oxford, pp. 9–20.

Hemmingsen, A. M. (1960). Energy metabolism as related to body size and respiratory surfaces and its evolution. *Rep. Steno. Mem. Hosp. Nord. Insulin lab.* **9**, 1–110.

Hill, A. V. (1950). The dimensions of animals and their muscular dynamics. *Sci. Prog., Oxf.* **38**, 209–230.

Hochaka, P. W. and Somero, G. N. (1973). "Strategies of Biochemical Adaptation". W. B. Saunders Co., Philadelphia, London, Toronto, 358 pp.

Holling, C. S. (1966). The strategy of building models of complex ecological systems. *In* "Systems Analysis in Ecology" (Ed. K. F. Watt). Academic Press, London and New York, pp. 195–214.

Hubbell, S. P. (1971). Of sowbugs and systems: the ecological energetics of a terrestrial isopod. *In* "Systems Analysis and Simulation in Ecology", vol. 1 (Ed. B. C. Patten). Academic Press, New York and London, pp. 269–323.

Ivlev, V. S. (1961). "Experimental Ecology of the Feeding of Fishes." Yale University Press, New Haven, 302 pp.

Jennings, J. B. (1957). Studies on feeding, digestion, and food storage in free-living flatworms (Platyhelminthes: Turbellaria). *Biol. Bull.* **112**, 63–80.

Jennings, J. B. and Calow, P. (1975). The relationship between high fecundity and the evolution of entoparasitism. *Oecologia* **21**, 109–115.

Johnson, L. (1966). Experimental determination of food consumption of pike, *Esox lucius*, for growth and maintenance. *J. Fish. Res. Bd. Can.* **23**, 1495-1505.

Kelso, J. R. M. (1972). Conversion, maintenance and assimilation for walleye, *Stizostedion vetreum vitreum*, as affected by size, diet and temperature. *J. Fish. Res. Bd. Can.* **29**, 1181–1192.

Kerr, S. R. (1971). Analysis of laboratory experiments on growth efficiency of fishes. *J. Fish. Res. Bd. Can.* **28**, 801-808.

Kleiber, M. (1961). "The Fire of Life." John Wiley and Sons Inc., New York, 454 pp.

Krogh, A. (1916). "The Respiratory Exchange of Animals and Man." Longmans and Green, London, 173 pp.

Kurtén, B. (1953). On the variation and population dynamics of fossil and recent mammal populations. *Acta. zool. fenn.* **76**, 1–122.

La Mettrie, J. O. de (1960). "L'homme machine. A Study in the origins of an Idea." Critical edition with an introductory monograph and notes by Aram Vartanian. Princeton University Press, Princeton, New Jersey, 269 pp.

Lawton, J. H. (1970). Feeding and food energy assimilation in larvae of the damselfly *Pyrrhosoma nymphula* (sulz.) (Odonata: Zygoptera). *J. Anim. Ecol.* **39**, 669–689.

León, J. A. (1976). Life histories as adaptive strategies. *J. theoret. Biol.* **60**, 301–336.

Lewontin, R. C. (1965). Selection for colonizing ability. *In* "The Genetics of Colonizing Species" (Eds. H. G. Baker and G. L. Stebbins). Academic Press, New York, pp. 79–94.

Loschiavo, S. R. (1968). Effect of oviposition on egg production and longevity in *Trogoderma parabile* (*Coleoptera: Dermestidae*). *Can. Ent.* **100**, 86–89.

Lotka, A. J. (1922). Contribution to the energetics of evolution. *Proc. natn. Acad. Sci. U.S.A.* **8**, 147–151.

Markus, H. C. (1934). Life history of the black headed minnow, *Pimephales promelas. Copeia* **70**, 116–122.

Maynard Smith, J. (1959). The rate of ageing in *Drosophila subobscura. In* "CIBA Foundation Colloquia on Ageing", vol. 5 (Eds. G. E. W. Wolstenholme and M. O'Connor). J. & A. Churchill Ltd., London, pp. 269–285.

Mayr, E. (1956). Geographic character gradients and climatic adaptation. *Evolution* **10**, 105–108.

Mayr, E. (1961). Cause and effect in biology. *Science, N.Y.* **134**, 1501–1506.

McClendon, J. H. (1975). Efficiency. *J. theor. Biol.* **49**, 213–218.

McLaren, I. A. (1966). Adaptive significance of large size and long life of the chaetognath, *Sagitta elegans*, in the Arctic. *Ecology* **47**, 852–855.

McNab, B. K. (1971). On the ecological significance of Bergmann's Rule. *Ecology* **52**, 845–854.

McNeill, S. and Lawton, J. H. (1970). Annual production and respiration in animal populations. *Nature (Lond.)* **225**, 472–474.

Medawar, P. B. (1952). "An Unsolved Problem in Biology". Lewis, London, 24 pp.

Miner, R. W. (1955). The regulation of hunger and appetite. *Ann. N.Y. Acad. Sci.* **63**, 1–144.

Murdoch, W. W. (1966). Population stability and life history phenomena. *Am. Nat.* **100**, 5–11.

Needham, A. E. (1964). "The Growth Processes in Animals." Sir Isaac Pitman and Sons Ltd., London, 522 pp.

Nevo, E. (1973). Adaptive variation in size of cricket frogs. *Ecology* **54**, 1271–1281.

Odum, E. P. (1971). "Fundamentals of Ecology". Saunders, Philadelphia and London, 574 pp.

Odum, E. P., Marshall, S. G. and Marples, T. G. (1963). The calorific content of migrating birds. *Ecology* **46**, 901–904.

Odum, H. T. and Pinkerton, R. (1955). Time's speed regulator. *Am. Sci.* **43**, 341–343.

Orgel, L. E. (1973). Ageing of clones of mammalian cells. *Nature (Lond.)* **243**, 441–445.

Osbourne, T. B. and Mendel, L. B. (1915). The suppression of growth and the capacity to grow. *J. Biol. Chem.* **18**, 95–106.

Paine, R. T. and Vadas, R. L. (1969). Calorific values of benthic marine algae and their postulated relation to invertebrate food preference. *Mar. Biol.* **4**, 79–86.

Paloheimo, J. E. and Dickie, L. M. (1965). Food and growth of fishes. I. A growth curve derived from experimental data. *J. Fish. Res. Bd. Can.* **22**, 521–542.

Paloheimo, J. E. and Dickie, L. M. (1966). Food and growth of fishes. III. Relations among food, body size and growth efficiency. *J. Fish. Res. Bd. Can.* **23**, 1209–1248.

Pandian, T. J. (1967). Transformation of food in the fish *Megalops cyprinoides*. *Mar. Biol.* **1**, 107–109.

Parnas, H. and Cohen, D. (1976). The optimal strategy for the metabolism of reserve materials in micro-organisms. *J. theor. Biol.* **56**, 19-55.

Pentelow, F. J. K. (1939). The relation between growth and food consumption in the brown trout *(Salmo trutta)*. *J. exp. Biol.* **16**, 446–473.

Peterson, B. (1950). The relation between size of mother and number of eggs and young in some spiders. *Experimentia* **6**, 96–98.

Prigogine, I. (1955). "Introduction to Thermodynamics of Irreversible Processes". Springfield, Illinois, 119 pp.

Prinslow, T. E. and Valiela, I. (1974). The effect of detritus and ration size on the growth of *Fundulus heteroclitus* (L). *J. exp. mar. Biol. Ecol.* **16**, 1–10.

Rahn, O. (1940). Efficiency of energy utilization in the growth of bacteria. *Growth* **4**, 77–80.

Reynoldson, T. B. (1966). The distribution and abundance of lake-dwelling triclads—towards a hypothesis. *In* "Advances in Ecological Research", vol. 3 (Ed. J. B. Cragg). Academic Press, London and New York, pp. 1–71.

Richardson, A. M. M. (1975). Energy flux in a natural population of the land snail *Cepaea nemoralis* L. *Oecologia* **19**, 141–164.

Roughgarden, J. (1971). Density dependent natural selection. *Ecology* **52**, 453–468.

Rubner, N. (1908). Probleme des Wachstums und der Lebensdauer. *Mitt. Ges. inn. Med. Wein Suppl.* **9**, 58–81.

Rubner, M. (1924). Aus dem Leben des Kaltblüters. *Biochem, Z.* **148**, 222–307.

Russell-Hunter, W. (1961). Life cycles of four freshwater snails in limited populations in Loch Lomond, with a discussion on intraspecific variation. *Proc. zool. Soc. Lond.* **137**, 135–171.

Russell-Hunter, W. D. (1970). "Aquatic Productivity". The Macmillan Company, London, 306 pp.

Salthe, S. N. (1969). Reproductive modes and the number and size of ova in the urodeles. *Am. Midl. Nat.* **81**, 467–490.

Schindler, D. W., Clark, A. S. and Gray J. R. (1971). Seasonal calorific values of freshwater zooplankton as determined with a Phillipson Bomb Calorimeter modified for small samples. *J. Fish. Res. Bd. Can.* **28**, 559–564.

Schindler, J. E. (1971). Food quality and zooplankton nutrition. *J. Anim. Ecol.* **40**, 589–595.

Schoener, J. W. and Janzen, D. H. (1968). Notes on environmental determinants of tropical versus temperate insect size patterns. *Am. Nat.* **102**, 207–224.

Scholander, P. F. (1955). Evolution of climatic adaptation in homeotherms. *Evolution* **9**, 15–26.

Schrödinger, E. (1944). "What is life?" Cambridge University Press, Cambridge, 91 pp.

Slobodkin, L. B. (1960). Ecological energy relationships at the population level. *Am. Nat.* **94**, 213–236.

Slobodkin, L. B. (1962). Energy in animal ecology. *In* "Advances in Ecological Research", vol. 1 (Ed. J. B. Cragg). Academic Press, New York, pp. 69–101.

Slobodkin, L. B. and Richman, S. (1961). Calories/gm in species of animals. *Nature (Lond).* **191**, 299.

Snedecor, G. W. (1956). "Statistical Methods", (fifth edition). Iowa State University Press, Iowa, 534 pp.

Southwood, T. R. S. (1976). Bionomic strategies and population parameters. *In* "Theoretical Ecology" (Ed. R. M. May). Blackwell Scientific Publications, Oxford, pp. 26–48.

Spiegelman, S. (1971). An approach to the experimental analysis of precellular evolution. *Q. Rev. Biophys.* **4**, 213–53.

Stanley, S. M. (1973). An explanation for Cope's rule. *Evolution* **27**, 1–26.

Teal, J. M. (1957). Community metabolism in a temperate cold spring. *Ecol. Monogr.* **23**, 41–78.

Tinkle, D. W. (1969). The concept of reproductive effort and its relation to the evolution of life histories of lizards. *Am. Nat.* **103**, 501–516.

Tinkle, D. W., Wilbur, H. M. and Tilley, S. G. (1970). Evolutionary strategies in lizard reproduction. *Evolution* **24**, 55–74.

Townsend, C. R. (1975). Strategic aspects of time allocation in the ecology of a freshwater pulmonate snail. *Oecologia* **19**, 105–115.

Warren, C. E. and Davis, G. E. (1967). Laboratory studies on the feeding, bioenergetics and growth of fish. *In* "The Biological Basis of Freshwater Fish Production" (Ed. S. D. Gerking). Blackwells, Oxford, pp. 175–224.

Weatherly, A. H. (1966). "Growth and Ecology of Fish Populations". Academic Press, London, 293 pp.

Weiss, P. and Kavanau, J. L. (1957). A model of growth and growth control in mathematical terms. *J. Gen. Physiol.* **41**, 1–47.

Welch, H. E. (1968). Relationship between assimilation efficiencies and growth efficiencies for aquatic consumers. *Ecology* **49**, 755–759.

Whitney, R. J. (1942). The relation of animal size to oxygen consumption in some fresh-water turbellarian worms. *J. exp. Biol.* **19**, 168–175.

Wiegert, R. G. and Coleman, D. C. (1970). Ecological significance of low oxygen consumption and high fat accumulation by *Nasutermes costalis* (Isoptera: Fermitidae). *Bioscience* **20**, 663–665.

Williams, G. C. (1957). Pleiotropy, natural selection and the evolution of senescence. *Evolution* **11**, 398–411.

Wilson, D. S. (1973). Food selection among copepods. *Ecology* **54**, 907–914.

Wilson, D. S. (1974). Prey capture and competition in the ant lion. *Biotropica* **6**, 187–193.

Wilson, D. S. (1975). The adequacy of body size as a niche difference. *Am. Nat.* **109**, 769–784

Wilson, P. N. and Osbourn, D. F. (1960). Compensatory growth after undernutrition in mammals and birds. *Biol. Rev.* **35**, 324–363.

Windell, J. T. (1966). Rate of digestion in the blue-gill sunfish. *Invest. Indiana Lakes and Streams* **7**, 185–217.

Wissing, T. E., Darnell, R. M., Ibrahim, M. A. and Berner, L. (1973). Calorific values of marine animals from the Gulf of Mexico. *Contributions in Marine Science* **17**, 1–7.

Wynne-Edwards, V. C. (1962). "Animal Dispersion". Oliver and Boyd, Edinburgh and London, 653 pp.

Young, E. C. (1965). Flight muscle polymorphism in British Corixidae: ecological observations. *J. Anim. Ecol.* **34**, 353–390.

Notes Added in Proof

The results from a number of studies which are of direct relevance to my arguments have appeared since this article went to press. Here I wish to draw attention to some of the more important of these findings.

1. A symposium on "Lipids in Animal Life Histories", with an introduction by Derickson (1976), underlines the point made in my Sections I–IV that the partitioning of energy between the several metabolic compartments, particularly between the storage and other compartments (Section VI), is likely to be adaptive and thus explicable in terms of present ecology and past evolutionary history (see also Calow and Jennings, 1977). Two theoretical papers in the symposium are particularly interesting. One by Smith (1976) draws attention to the trade-off between speed and efficiency in metabolism and points to a general switch from speed to efficiency in the course of succession (see my Section IV). The other paper, by Pianka (1976), begins a theoretical attack on the "egg size *v* numbers problem" (see also my Section VII).

2. Maiorana (1976) in a paper on salamanders has found that flexibility in partitioning energy between reproduction and other aspects of metabolism

is important in the ecology of the plethiodontid species. She argues, as I do (Sections XII and XIII), that the degree of control exercised on reproductive output may be important in determining adult life-span. Confirming this interpretation, we (Calow and Woollhead, 1977a) have found that during nutritive stress reproductive effort rises dramatically in semelparous but not iteroparous triclads.

3. A recent study on the metabolism, movement and mucus loss in triclads (Calow and Woollhead, 1977b) has confirmed the general conclusion on searching strategies presented in Section XI, A, a but has suggested that mucus loss, though a significant percentage of the energy ingested (20%), is less than the 70% quoted in Section IX.

4. A paper by Ellis *et al.* (1976) has presented a model which explains diet selection as a dynamic process dependent on both the internal state of the feeder and external conditions (see my Section XI, B).

References

Calow, P. and Jennings, J. B. (1977). Optimal strategies for the metabolism of reserve materials in microbes and metazoa. *J. theoret. Biol.* **66**, in press.

Calow, P. and Woollhead, A. S. (1977a). The relationship between ration, reproductive effort and age-specific mortality in the evolution of life-history strategies—some observations on freshwater triclads. *J. Anim. Ecol.* (in press).

Calow, P. and Woollhead, A. S. (1977b). Locomotory strategies in freshwater triclads and their effects on the energetics of degrowth. *Oecologia* **27**, 353–362.

Derickson, K. W. (1976). Introduction to the symposium on lipids in animal life histories. *Am. Zool.* **16**, 629–630.

Ellis, J. E., Wiens, J. A., Rodell, C. Z. and Anway, J. C. (1976). A conceptual model of diet selection as an ecosystem process. *J. theoret. Biol.* **60**, 93–108.

Maiorana, V. C. (1976). Size and environmental predictability for salamanders. *Evolution* **30**, 599–613.

Pianka, E. R. (1976). Natural selection of optimal reproductive tactics. *Amer. Zool.* **16**, 775–784.

Smith, C. (1976). When and how much to reproduce: the trade-off between power and efficiency. *Amer. Zool.* **16**, 763–774.

Rodent Long Distance Orientation ("Homing")

JAMES K. JOSLIN

*Department of Biology, George Mason University,
Fairfax, Virginia*

I. INTRODUCTION

Mammalian orientation has been most extensively studied in bats and rodents. There is a comprehensive review of bat homing studies (Davis, 1966), but there is no comparable review of long distance orientation ("homing") for rodents. With the major exception of rat maze studies, most rodent orientation studies have dealt with homing orientation rather than with orientation within the home range (short distance orientation). This review will concentrate on rodent homing studies but will include some short distance orientation studies which may be pertinent to a discussion of homing. For recent information on orientation by various taxa of fish, amphibia, reptiles, birds, and mammals, the reader should refer to the numerous articles in Storm (1967), Orr (1970), Adler (1971), and Galler *et al.* (1972); for various invertebrate taxa, Orr (1970), Adler (1971), and Galler *et al.* (1972); and for a recent review paper on bird orientation and navigation, Keeton (1974).

The purposes of this review are to summarize information on rodent homing orientation and to demonstrate that it is more likely to be based on piloting (orienting by using *familiar* cues in the environment) than on random search or sophisticated navigational abilities. I shall consider methodological aspects of rodent homing studies, possible adaptive values of rodent long distance orientation, factors influencing this orientation, and hypotheses regarding its mechanisms. Finally, I shall conclude with suggestions for manipulative studies to critically test the familiarity hypothesis.

II. Methodological Aspects

A. definitions

Most homing studies have involved releases of rodents at distances several times greater than their home range diameters. Home range is that area within which the animal spends most of its time. Homing (long distance orientation) is considered to have taken place if the rodent is subsequently observed (usually via live trap capture) to have returned to its home range (Stickel, 1968). Generally, home range is determined by at least three captures of an individual before it is experimentally displaced. The home range is part of an individual's life range, which also includes other areas the rodent encounters on its own during its lifetime.

Short distance orientation occurs within a rodent's home range. Movement patterns here are commonly studied by noting the relationships of tracks to various objects (Stickel, 1968) or by directly observing the rodent on live trap release (Smith and Speller, 1970). Short distance orientation is also present in laboratory studies of rats' and other rodents' maze running.

B. homing displacement methods

1. *Transport and release*

In homing studies, rodents are released usually 25–2000 m beyond their home ranges (or live trap last captured in), after having been marked for individual recognition. Commonly, individuals are recognized by clipped toes, ear tags, holes punched in ears, or a combination of such methods. Layne (1957) has used clipped fur patches.

An individual is transported to the displacement site usually in an opaque container; unfortunately, in some studies (Robinson and Falls, 1965; Furrer, 1973) the rodent could see out. Experimenters rarely attempt to eliminate possible kinesthetic cues by taking an indirect route or by using a rotating turntable in transit. Robinson and Falls (1965) and De Busk and Kennerly (1975), however, have made such attempts. Experimenters release a rodent by removing the container or by allowing the rodent to exit. Either way, they do not attempt to initially orient the rodent in any given direction. Sometimes individuals are released jointly, rather than singly, at the same location (Murie, 1963; Gentry, 1964). Such a release method is undesirable, for it might introduce confounding social factors. Furrer (1973) suggested that these could include agonistic encounters (which might disorient some individuals) or several mice following a more experienced mouse. Various experimenters (Gentry, 1964; Fisler, 1966; Furrer, 1973) have released their marked individuals during the daytime, even though they belong to a nocturnal species. This might be a poor procedure for it could increase mortality, and perhaps even lessen a motivation to home the following night. It could also introduce additional error into calculating homing time, since a nocturnal species is likely to seek nearby shelter (Gentry, 1964; Furrer, 1973) and not start homing until many hours later.

2. Direction traveled after release

Several methods based on movements close to the release site have been used to indicate a homing tendency. In the laboratory, a rodent was released in the center of a labyrinth maze; its position at the periphery was assumed to indicate directional preference (Lindenlaub, 1955; Bovet, 1960). In the field, Fisler (1967) used a similar technique with a confined circular platform. The initial observable parts of a rodent's route upon release in a live trap recapture study have also been used as indicators of homing. Griffo (1961) marked deermice with luminescent tape and directly observed them after release at night on a golf course, an area where they would be readily visible. Snow tracking is more accurate but special conditions are necessary: fresh snow, no further snow, and no wind. The possibility of a rodent using the experimenter's tracks for orienting can be minimized by making a symmetrical pattern of human tracks leading from the release site (Bovet, 1968).

Live trap release and recapture studies yield scant information on the route taken during homing or on the real homing speed (Rawson and Hartline, 1964). However, this is the method most frequently used in homing studies because it does not require special conditions or sophisticated technology, and its costs are minimal. But, potentially, an experimenter could use live trap release and recapture to compare the relative importance of directions to the home site, home site locations, and release site locations. Saint Girons and Durup (1974) have suggested a square geometric design for these comparisons, with the two release points and the two home sites at alternate corners. Thus, the three factors can be separated for analysis, and potential homing distances can be held constant.

Radiotelemetry can describe a rodent's *complete* homing route and can more accurately measure homing speed (Rawson and Hartline, 1964). But loss of radio contact and small sample sizes are disadvantages. Possibly because of cost and technical problems, this technique has rarely been used in rodent homing studies.

3. *Percentage of subjects homing and displacement distance*

The basic dependent variables in live trap recapture homing studies are the percentage of subjects homing when displaced at various distances from the home range, and homing time.

Typically, the percentage of subjects homing is inversely related to the displacement distance. This has been demonstrated in field studies of *Apodemus sylvaticus*, *A. flavicollis*, and *Clethrionomys glareolus* (Lehmann, 1956; Bovet, 1962); *Peromyscus leucopus* (Burt, 1940), *P. maniculatus* (Murie and Murie, 1931; Murie, 1963; Bovet, 1972; Furrer, 1973), *P. gossypinus* (Griffo, 1961); *Microtus pennsylvanicus* (Robinson and Falls, 1965); *Reithrodontomys megalotis* (Fisler, 1966); *Tamiascurus hudsonicus* (Layne, 1954); *Ondatra zibethica* (Mallach, 1972); and *Sigmodon hispidus* (De Busk and Kennerly, 1975). Rawson (1966) claimed no relationship between distance and homing success for *P. maniculatus*, but he provided no data. Displacement distances in these studies were usually somewhere within a range of 70–1600 m, with percentage homing generally varying between 85% and 0%.

Workers have offered diverse hypotheses on the mechanisms underlying this inverse relationship between homing success and displacement distance. The most frequent (and plausible) hypothesis asserts varying degrees of familiarity with areas around the home nesting site. Griffo (1961) and Furrer (1973), for instance, have advanced a familiarity hypothesis to explain a gradual decrease in the percentage of *Peromyscus* homing for the first several hundred meters, with a rapid decrease after 650 m. The gradual initial decrease might be due to the deermice's partial familiarity with an area (an average life range);

the rapid decrease after 650 m might reflect lack of familiarity with an area near the release site, particularly for younger deermice. In further support of this familiarity hypothesis, Furrer (1973) found significant differences in percentage return of experienced and inexperienced deermice only for deermice displaced more than 650 m. For other rodent studies, a non-linear relationship between displacement distances and percentage of subjects homing cannot readily be ascertained because of the restricted range of displacement distances and small sample sizes.

A familiarity hypothesis has also been applied to homing for other rodent genera. For example, Robinson and Falls (1965) reasonably assert that at short distances *Microtus pennsylvanicus* has complete familiarity with the release area; at intermediate distances it has less familiarity (fewer voles know the area and they recognize fewer landmarks); and at long distances it has no familiarity with the release area. Their assertions are supported by the poorer homing success of enclosure-reared voles compared with controls when both were field-released at the same distance, and by their survey of the homing literature. This survey revealed that the maximal displacement distance from which a rodent homed was directly related to the home range size.

Differential familiarity with an area (life range) and observed percentage homing success have also been explained by the home range being a series of concentric probability-of-occurrence zones (Robinson and Falls, 1965); the home range being under-estimated by various live-trapping methods (Chitty, 1937; Robinson and Falls, 1965); the shifting of home range by an individual (Griffo, 1961); exploratory trips outside the home range (Stickel and Warbach, 1960; Griffo, 1961; Furrer, 1973); and by the distribution of juveniles' dispersal distances outside the parents' home ranges (Griffo, 1961).

Could random search account for the inverse relationship between percentage homing and displacement distance? Robinson and Falls (1965) have examined this possibility more closely than have other authors. They found that several different random search models failed to help them interpret their data.

Authors have advanced other explanations for the inverse relationship besides differential familiarity or random search. Hamilton (1937) and Bovet (1962) speculated that a rodent may have increased opportunity for encountering favorable and (supposedly) unoccupied habitat farther from the home site. Thus, it might establish a new home. Bovet (1962) also speculated that increased probability of individual mortality through predation may occur as displacement distance increases. However, Robinson and Falls (1965) noted that practically all of their voles returned on second releases, regardless of displacement distance.

Bovet (1962) has also assumed that a constant angular error of orientation may be present, which for greater distances would result in a rodent having less chance to encounter its home range.

4. *Homing times*

For a given distance, homing times typically have been highly variable. Time usually has been measured in whole day or night increments. Numerous studies illustrate this variability. For example, Fisler (1962) recorded a range of 1–9 days for 200 ft displacements in voles; Murie (1963), 1–8 nights for 800 yd displacements in deermice; and Fisler (1966), 1–47 days for 320 ft displacements in harvest mice. Fisler (1962) has suggested that in *Microtus californicus* these individual differences could be attributed to physiological conditions, social and psychological motives, wandering tendencies, and dispersal tendencies. Some of this individual variability could be caused by individual differences in familiarity with the surrounding area (Robinson and Falls, 1965). Also, individuals may enter live traps at varying times after they have returned to their home ranges. Radiotelemetric tracking of *Peromyscus maniculatus* (Rawson and Hartline, 1964) suggests this possibility. For example, one of their mice did not enter a live trap the first night after it had returned home.

Homing time is not related to displacement distance (Schleidt, 1951; Fisler, 1962, 1966; Murie, 1963; Robinson and Falls, 1965). One reason may be the sources of variability between individuals. Another may be the way time has been measured. Rawson and Hartline (1964) illustrate the crudeness of the usual time measurement. Thus, for *Peromyscus maniculatus*, in live trap studies (the usual method) homing speed is about 30 m/hr; however, using radiotelemetry Rawson and Hartline (1964) obtained a homing speed of 300 m/hr. But crudeness of the usual time measurement alone probably cannot account for the lack of a relationship between homing time and distance displaced. A considerable degree of individual variability in return time also existed in a study in which live traps were checked at two-hour intervals (Gentry, 1964).

III. ADAPTIVE VALUES OF LONG DISTANCE ORIENTATION

For this speculative discussion, I will use *Peromyscus* (deermice) to illustrate the possible adaptive values of rodent long distance orientation. I will discuss these adaptive values in terms of the three factors which influence a rodent's life range size: juvenile dispersal, occasional adult wandering, and shifting of the home range (Griffo, 1961). Orientation within the home range will also be considered, for there is no reason to believe that *Peromyscus* would not find similar advantages in orienting over both longer and shorter distances.

Juvenile dispersal, one factor influencing life range size, might be adaptive by spacing out individuals, thereby helping to prevent overpopulation (Terman, 1968). *Peromyscus* juveniles often exhibit short distance dispersal (Terman, 1968); the areas traversed become part of the individuals' life ranges (Stickel, 1968). Subsequent orientation within these areas would be enhanced if the deermice had good memories. Griffo (1961) demonstrated a good memory for *P. gossypinus*, which homed readily even after five to twelve weeks in captivity. The wandering tendency present during juvenile dispersal is perhaps retained in adulthood. This might account, at least partially, for occasional adult wandering ("exploratory trips").

Peromyscus adult exploratory trips have been demonstrated in field studies (Stickel, 1968). Their tendencies to explore without food, water, or other obvious rewards have been demonstrated in the laboratory (Brant and Kavanau, 1964, 1965). These exploratory trips, which would help the deermice establish their life ranges, may be adaptive in terms of their seeking a mate and in promoting various aspects of

TABLE I

Average homing distances of two Peromyscus *species. The homing distances, much greater than the average home range radius, are related to the average juvenile and adult dispersal components of the species' life ranges. These species are apparently familiar with an area considerably larger than their home range. Data derived from Stickel (1968).*

	Home range radius (m)	Juvenile dispersal from nest site (m)	Adult dispersal from home range center (m)	Homing distance (m)
P. leucopus noveboracensis	19·2	190·6	—	170·1
P. maniculatus bairdi	23·9	159·6	281·1	112·8

their social organization, such as a dominance hierarchy (Stickel, 1968). Adult and juvenile dispersals could result in deermice being familiar with an area much larger than the home range. Table I suggests that these dispersals may help account for deermice homing from displacement distances far outside the home range.

Maintenance of the social organization and of the distribution of individuals seems to be an adaptive value of both exploratory trips

and orientation within the home range. However, more evidence exists
for its adaptiveness within the home range. *P. maniculatus*, released
some distance from their nests, returned to them and avoided using
empty nests of their neighbors (Terman, 1962). Even when some *P.
leucopus* were removed from an island, other individuals did not
quickly expand their home ranges (Sheppe, 1966). Attachment to the
home range may help maintain an individual's status in the social
structure and confer an advantage over aliens who enter this area
(Sheppe, 1966). This advantage has been directly observed in *P.
polionotus*. Several instances of residents evicting live-trap-released
transients from their refuge holes were noted by Blair (1951).

Prior exploratory trips might also enhance *Peromyscus* individuals'
survival and subsequent return to their home ranges, after having been
temporarily forced from them by natural events such as drought,
flood, fire (Stickel, 1968), predators (Murie and Murie, 1931), or com-
peting species (Grant, 1971).

In some cases, *Peromyscus* individuals may permanently shift their
home ranges in response to such relatively rare natural events (Stickel,
1968). Prior familiarization with areas outside their previous home
ranges might enhance survival in the new home ranges. Adult *Peromyscus* also shift their home ranges in response to a change in the distribution of preferred foods (Stickel, 1968) or in response to artificial
provisioning of food (Sheppe, 1966). Here, too, exploratory trips might
precede home range shifting.

Peromyscus locates food sources by orienting on exploratory trips
and within its home range. Again, there is more evidence in terms of
home range orientation. On release from live traps within its home
range, *P. maniculatus* moved directly toward a food cache and subsequently dug for the food (McCabe and Blanchard, 1950). The
frequency of *P. leucopus* travels to various parts of its home range
changed according to the location of food sources (Sheppe, 1966). *P.
polionotus* has heavily-frequented trails leading directly to areas where
sea oats were blown down (Blair, 1951). The extent of *Peromyscus*'
orienting movements may be a function of food availability, with
smaller home ranges when the food supply is abundant (Stickel, 1968).

Efficient avoidance of predators is another function of home range
and perhaps of life range orientation. After live trap release within
the home range, *P. maniculatus* frequently sought holes at tree bases
(Smith and Speller, 1970); other *Peromyscus* species also oriented to
holes as refuge sites (McCabe and Blanchard, 1950; Stickel, 1968).
Some laboratory evidence also suggests that home range orientation
could enhance avoidance of a predator. In a large room, resident *P.
leucopus* or individuals unfamiliar with the room ("transient" deermice)

were confined separately with an owl. A significantly higher precentage of the residents survived (Metzgar, 1967).

IV. Factors influencing Homing

A. extrinsic factors

1. *Physical and biotic factors*

A rodent's homing performance is influenced by physical and biotic environmental factors. Burt (1940) observed a southward bias in movements for *Peromyscus leucopus*; Bovet (1962), a southward and to a lesser extent an eastward and westward bias for *Apodemus flavicollis* and *A. sylvaticus*; and Saint Girons and Durup (1974), a northward and southward bias for *Clethrionomys glareolus*. Predation or the location of the individuals' home sites did not readily account for these differences in live trap recaptures. Perhaps differential landmark familiarity and habitat heterogeneity could partially account for these directional movement biases. Bovet (1962) concluded that the observed directional biases implied that his mice were orienting "according to [unspecified] factors they perceived at the release site". These factors may well be based on habitat heterogeneity. In all three of these studies there was considerable habitat heterogeneity, perhaps enough to bias movement patterns. Directional differences in bat homing have been interpreted in terms of differential familiarity with landmarks (Davis, 1966); this could be based on habitat heterogeneity.

There is evidence that type of habitat affects rodent homing success. Bovet (1965a) observed that *Apodemus sylvaticus* homed with greater success from a zone relatively poor in vegetation and topographical features than from one rich in those features. Effects of habitat on homing success have also been noted for *Peromyscus* (Griffo, 1961; Murie, 1963); for *Microtus* (Fisler, 1962; Robinson and Falls, 1965); for *Reithrodontomys* (Fisler, 1966); and *Neotoma* (Lay and Baker, 1938).

There are diverse interpretations for the effects of habitat on homing success. Heterogeneous habitat provides distinctive cues that could facilitate homing (Murie and Murie, 1931), while poor habitat, which lacks sufficient basic biological requirements of the species such as food, cover or nest sites provides few distinctive cues for orienting (Griffo, 1961; Robinson and Falls, 1965). In some cases, poor habitat at a release site offers insufficient concealment, enhancing the rodent's motivation to return home (Fisler, 1966; De Busk and Kennerly, 1975). Poor habitat in one direction from the release site reduces movement in that direction (Murie, 1963; Furrer, 1973; Savidge, 1973; Saint Girons and Durup, 1974); depending on the home site location, this could either increase or decrease homing success. All of these interpretations are consistent with (or at least do not preclude) the possibility of a rodent's homing by using familiar cues. Experiments have not been designed specifically to discriminate between these *post hoc* interpretations.

Seasonal effects have been implicated for *Apodemus sylvaticus*, which homes better in winter than in summer. Winter subjects may have been more familiar with surrounding areas and may also have been more likely to return home rather than establish a new home site (Lehmann, 1956). Other factors may also be involved, for this species exhibited better initial orientation (homing direction) in a laboratory maze in winter than in summer (Bovet, 1960). However, homing by *Apodemus sylvaticus* was very poor on the coldest winter nights, perhaps because of increased mortality (Saint Girons and Durup, 1974). This winter effect may be species specific, for *Mus musculus* was disoriented in its initial maze movements in winter (Lindenlaub, 1955).

2. *Interactions of individuals*

Intraspecific social interactions have been implicated more often than interspecific interactions as factors affecting homing. Social interactions with conspecifics near the release site can facilitate homing, perhaps by increasing the stress a rodent experiences (Griffo, 1961; Fisler, 1962; De Busk and Kennerly, 1975). This could explain why displaced *Microtus californicus* remained near release sites that lacked resident voles but moved away from release sites that had resident voles (Fisler, 1962). *Reithrodentomys megalotis*, however, did not show such differences (Fisler, 1966). Social repulsion might occur as a result of direct aggression between individuals. Fisler (1962) had suggested this for *Microtus californicus*, and De Busk and Kennerly (1975) for *Sigmodon hispidus*. Or, social repulsion might result from a tendency of individuals to mutually avoid each other. Terman (1962) and Murie (1963) have suggested this type of repulsion for *Peromyscus*.

However, interactions with conspecifics are not essential for homing to occur. Stickel (1949) found that most *Peromyscus* homed even though they were moving through areas in which all conspecifics had been experimentally removed. Indeed, Bovet (1965a, 1968) has suggested that homing by *Apodemus sylvaticus* and by *Peromyscus maniculatus* released in winter was enhanced by a lack of diverting cues such as conspecifics, food, or cover. This hypothesis does not necessarily exclude the social repulsion hypothesis. Perhaps in some instances an individual's reaction to social repulsion might be so extreme as to lead to disorientation (being diverted) rather than to successful homing.

At the home site, the presence of a conspecific can *inhibit* homing.

Terman (1962) tethered a young alien *Peromyscus maniculatus* by the nest box of each adult deermouse. He found that adults then homed significantly less to their nest boxes, perhaps because individuals tend to avoid each other. The presence of conspecifics at the newly vacated home site of *Clethrionomys glareolus* and *Apodemus sylvaticus* probably also accounts for their poor homing when retested after a few days in captivity; this seems particularly likely since there were high population densities for both species (Saint Girons and Durup, 1974).

Homing may also be adversely affected by interspecific interactions. *P. maniculatus* became less sedentary (percentage homing success decreased) about the same time that the first active ground squirrels were observed around the granaries occupied by *Peromyscus* (Bovet, 1970).

B. INTRINSIC FACTORS

1. *Practice*

Intrinsic factors (characteristics of individuals) have received greater emphasis in homing studies than have extrinsic factors. Various authors have noted improvement of individual homing performance on subsequent releases from the same location. Measures of improvement were decreased homing time (Griffo, 1961; Murie, 1963; Durup et al., 1973; Saint Girons and Durup, 1974), increased percentage of homing individuals (Robinson and Falls, 1965; Furrer, 1973; Saint Girons and Durup, 1974), and faster movement away from the release point (Griffo, 1961). However, Bovet (1968) found no improvement in the initial orientation of *Peromyscus maniculatus* on subsequent release, perhaps because of a very small sample size for second releases.

When subsequent releases were made at increasing distances but in the same direction from the home site, an improvement in performance often occurred as a decrease (or no change) in total homing time (Murie and Murie, 1931; Durup et al., 1973; Furrer, 1973). The percentage of subjects homing might also increase under these conditions (Robinson and Falls, 1965).

Various authors have also noted improvement of individual homing on subsequent releases in different directions. Here, rodent homing may improve in terms of decreased homing time (Griffo, 1961; Fisler, 1962; Furrer, 1973; Saint Girons and Durup, 1974), and increased percentage of homing individuals (Murie, 1963; Robinson and Falls, 1965; Saint Girons and Durup, 1974).

Improvement in homing on subsequent releases is probably due to practice effects; that is, a rodent acquires some familiarity with a large area during its first release movements (Griffo, 1961; Murie, 1963; Robinson and Falls, 1965). Two studies demonstrate the good memory that would be necessary. *Peromyscus gossypinus*, retained in the laboratory for five to twelve weeks after some homing experience, showed no subsequent decrease in homing ability when released at prior release sites (Griffo, 1961). *P. maniculatus*, although exhibiting no significant initial homeward orientation, did show a highly significant tendency to follow the same initial route on second displacement (Murie, 1963). However, Saint Girons and Durup (1974) obtained very poor subsequent homing after retaining *Apodemus sylvaticus* and *Clethrionomys glareolus* in captivity for only 36–84 hours. Unfortunately, their sample sizes were very small, and poor returns could be due to these rodents' vacated home sites being quickly occupied by conspecifics. This seems likely since these species' populations were at high densities. Saint Girons and Durup (1974) reasonably speculated that the new residents might then repel the original occupants.

Practice effects, however, are not the only explanation for improvement in homing on subsequent release. Furrer (1973) has suggested that individuals that home poorly on the initial release would not be available for subsequent releases, thus leaving a biased sample. Saint Girons and Durup (1974) have suggested that faster subsequent homing returns may represent a general adaptation of individuals to such experimental procedures as transport and release. There are apparently no data to support such speculations.

However, there are data from several studies which persuasively support a practice effect rather than these other possibilities. Saint Girons and Durup (1974) found that individuals released repetitively from the *same* site returned faster than did those released repeatedly from varying points but at the same distance from home. Robinson and Falls (1965) noted that voles retained in two-acre enclosures were less likely to return home after release outside the enclosure than were voles in unconfined neighboring areas. Only the unconfined voles would have had opportunities to become familiar with a large area. Robinson and Falls (1965) also noted that voles successfully homing from displacement distances of 200–1000 ft were unable to home successfully from 1110–1700 ft; their initial orientation was random. These results imply a lack of a directional sense on the part of these successfully homing individuals, and imply that they had acquired prior familiarity with distances of 1000 ft or less. Similarly, Furrer (1973) found that differences in percentage returns of experienced and inexperienced mice became highly statistically significant at the same distance where other data had indicated a sharp decline in homing ability for inexperienced mice. This suggests that both groups were equally familiar with areas near the home site, and that the experienced group had gained familiarity with a larger area as a result of prior releases.

2. *Age and stress*

In various studies, the percentage of adults that homed has generally been greater than that of juveniles. These include studies of *Peromyscus maniculatus* (Murie, 1963; Furrer, 1973), *P. gossypinus* (Griffo, 1961), *Microtus pennsylvanicus* (Robinson and Falls, 1965), *Reithrodontomys megalotis* (Fisler, 1966), and *Apodemus sylvaticus* and *Clethrionomys glareolus* (Durup *et al.*, 1973). Griffo (1961) has suggested this difference could occur because juveniles, not yet having dispersed, would have less familiarity with surrounding areas. Also, Griffo (1961) and Murie (1963) have attributed this to juveniles' lacking the motivation to return, perhaps because they have not yet established a home range of their own.

Murie (1963) presented some evidence that juveniles were less motivated to return. Using live traps arranged in a circle around the release point, he found that juveniles were more likely than adults to remain near the release point on subsequent nights. An alternative explanation, however, for these results is that after displacement, juveniles are less likely than adults to enter traps outside the home range. Although no specific age-related effect has apparently been demonstrated, some rodents do not readily enter live traps outside their home ranges during homing displacement experiments. This has been demonstrated for *Peromyscus gossypinus* (Griffo, 1961), *P. maniculatus* (Morris, 1967), and *Microtus pennsylvanicus* (Robinson and Falls, 1965).

Griffo (1961) has interpreted this decrease in capture outside the home range as being related to psychological and perhaps physiological stresses, which result in a search for the familiar area and an avoidance of live traps. Within the home range, the animal's "psychic" needs are met, it moves with "assurance," is not under stress, and enters live traps. However, there is no direct evidence for such postulated stresses on subjects displaced outside their home ranges.

Furthermore, either some rodents were not particularly stressed in this manner by displacement or they recovered rather quickly. Thus, rather than homing, displaced *adults* may establish *new* home ranges, often near the release site. This has been observed for *Peromyscus leucopus* (Stickel, 1949), *P. gossypinus* (Griffo, 1961), *P. maniculatus* (Murie, 1963; Rawson and Hartline, 1964), *P. polionotus* (Gentry, 1964), *Microtus pennsylvanicus* (Hamilton, 1937; Robinson and Falls, 1965), and perhaps *Reithrodontomys megalotis* of unspecified age (Fisler, 1966). Subsequent releases of such individuals have resulted in their returning to the "new" home rather than the "old" one (Griffo, 1961; Murie, 1963; Fisler, 1966). However, there is another interpretation for repeatedly capturing a rodent near the release point. Fisler (1966) has suggested that for *Reithrodontomys* the home range shifts frequently enough that a harvest mouse might be inadvertently released in its *previous* home range and remain there.

The displacement stresses postulated by Griffo (1961), if they occur, could well operate during only part of a year. In *Apodemus sylvaticus*, percentage homing success decreased as the mice became less sedentary (Bovet, 1965a). A similar effect, perhaps related to interspecific interactions, has been noted for *Peromyscus maniculatus* (Bovet, 1970).

3. Sex

Male rodents frequently exhibit a higher percentage homing success than females. This has been shown for most (or all) displacement distances for *Peromyscus leucopus* (Stickel, 1949), *P. maniculatus* (Murie, 1963; Furrer, 1973), *P. gossypinus* released from a natural habitat (Griffo, 1961), *Microtus californicus* (Fisler, 1962), and *M. pennsylvanicus* (Robinson and Falls, 1965). However, sex effects were minimal for *P. polionotus* (Gentry, 1964), and *Sigmodon hispidus* (De Busk and Kennerly, 1975); female *Apodemus sylvaticus* and *Clethrionomys glareolus* had a higher percentage homing success, perhaps because these species' females are more sedentary than males (Durup *et al.*, 1973). Pregnancy did not interfere with female homing for *P. maniculatus* (Murie, 1963) and *Reithrodontomys megalotis* (Fisler, 1966).

Perhaps male rodents usually home more successfully than females because the males may be familiar with a larger area. Stickel (1949, 1968) and Griffo (1961) cite evidence that males had larger home ranges; Griffo (1961) and Murie (1963), that males made longer exploratory trips outside their home ranges; and Griffo (1961), that juvenile males dispersed to longer distances. The differential familiarity hypothesis could be tested by comparing the homing of the two sexes over comparable distances in unfamiliar *versus* familiar habitats. Griffo (1961) obtained no sex differences in percentage homing for deermice released on a golf course (unfamiliar habitat), but claimed that his males homed better than females when released on natural habitats, which may have been more familiar to the males. Unfortunately, Griffo's claims were based on graphical comparisons of rather small sample sizes, not on statistical analysis of the data.

V. Hypotheses About Homing Mechanisms

A. Random Search

Most workers have drawn their conclusions on the random or non-random nature of homing from home site recapture and intermediate orientation data. Many have concluded that their rodents' homing performance was totally non-random. These include studies on *Peromyscus leucopus* (Burt, 1940), *P. maniculatus* (Rawson, 1966; Bovet, 1968), *Apodemus sylvaticus* (Bovet, 1962), and *Reithrodontomys megalotis* (Fisler, 1966). Other workers have concluded that their rodents' orientation was at least partially non-random. These include studies on *P. gossypinus* (Griffo, 1961), *P. maniculatus* (Murie and Murie, 1931; Morris, 1967; Furrer, 1973), *Clethrionomys glareolus* (Schleidt, 1951), *A. sylvaticus* and *A. flavicollis* (Lehmann, 1956), *Microtus pennsylvanicus* (Robinson and Falls, 1965), *M. californicus* (Fisler, 1962), *Tamiasciurus hudsonicus* (Layne, 1954), and *Tamias striatus* (Layne, 1957). The minority view of homing being due simply to random search has been proposed by Murie (1963).

Authors have usually based their conclusions on homing speeds and percentage homing success. Unfortunately, with the exception of Robinson and Falls (1965), their conclusions are not based on a comparison with any detailed model of randomness of movements. Thus, their conclusions (although probably reasonable) are rather subjective. Murie's conclusion of complete randomness apparently was based mostly on a lack of initial homeward orientation by his mice. This could be explicable on the basis of cover-seeking tendencies being predominant (Bovet, 1968).

The evidence from most studies seems to imply that homing movements are at least partially non-random. Direct snow tracking data (Bovet, 1968, 1971) and comparison of performances in natural and unnatural habitats (Griffo, 1961) are particularly impressive evidence in favor of a non-random component. Improvement in homing performance on subsequent release also implicates a non-random component. Homing from various displacement distances based completely on random search would be most likely for a species having very small home ranges, extremely slight tendencies to wander outside the home ranges, very small dispersal distances, and extremely poor distance receptors. Not many rodent species could meet all these criteria!

B. PILOTING AND SENSORY CUES

What mechanisms might be involved in non-random homing movements? Systematic search (for example, in the form of a spiral) combined with piloting (directed movements using cues present in a familiar area) has been suggested for some birds by Griffin and Hock (1949). The evidence for such a mechanism in birds is scant (Schmidt-Koenig, 1965); there is apparently no evidence for this in rodents. *A priori*, such a mechanism seems unlikely since it would probably require a high degree of orientational ability. A more goal-directed orientational mechanism would probably be selected for during the course of evolution.

A highly plausible combination of random search outside a familiar area and piloting in familiar areas has been frequently suggested as a mechanism underlying rodent orientation. Some aspects of this combination have already been discussed in the Random Search section of this paper. We will now consider the sensory cues utilized in piloting. Because of difficulties in directly observing a rodent's entire route while homing and in experimentally manipulating sensory cues *en route*, the evidence regarding sensory cues leaves much to be desired. However, the evidence, scant and imperfect as it is, does imply that the visual modality may play a major rôle in piloting.

1. *Visual cues*

Visual cues provided by the terrain are probably used in piloting. Practice effects (reviewed earlier) could imply use of such cues, as could the lower percentage homing success of enclosure-reared *Microtus pennsylvanicus* on release outside the enclosure (Robinson and Falls, 1965). However, *M. californicus* (and *Reithrodontomys megalotis*) did not seem to use the one class of terrain cues that was experimentally investigated. They did not exhibit initial homing orientation from a circular platform in the presence of horizon cues (Fisler, 1967). This negative evidence might be due, however, to the small size of the platform (1·65 m radius) and to lack of cover.

Considerable evidence indicates that, in piloting *Peromyscus* uses terrain visual cues (ground-associated, non-celestial cues). One *P. gossypinus* may have oriented to the heavy shadows of a wooded section near its release site on a golf course (Griffo, 1961). *P. leucopus* released 175 ft offshore from an island swam toward the island, perhaps orienting to tree tops (Sheppe, 1965a). *P. leucopus'* tracks on smoked paper in artificial shelters increased when empty juice cans, presumably visual cues, were placed near the shelters (Sheppe, 1965b). *P. maniculatus* usually ran directly to a nearby tree when released from a live trap (Smith and Speller, 1970). The possible importance of trees as piloting cues is also suggested by a laboratory study of *P. leucopus'* orientational cue preferences, a black vertical tube being most preferred, and by snow tracking observations (Joslin, 1971). In contrast, citing indirect evidence, Metzgar (1973) has concluded that this species does not orient to trees. He found that for *P. leucopus* the frequency of occurrence in unit areas of the home range decreased normally at greater distances from the center of activity, whereas large trees were purportedly scattered irregularly and widely over the study plot. It seems more probable, on the basis of the more direct types of evidence of Smith and Speller (1970) and Joslin (1971) that *P. leucopus* and other *Peromyscus* species might use conspicuous objects for piloting. Studies of *Peromyscus'* visual capacities support this viewpoint. A characteristic feature of this genus is its large eyes, which could imply enhanced visual capabilities (King, 1968). Indeed, *Peromyscus* has better visual acuity and visual range than other rodent species previously tested and has rapid visual maturation (Vestal, 1973; King and Vestal, 1974).

However, direct manipulation of *Peromyscus* sensory modalities, although not necessarily refuting the importance of visual stimuli in piloting, does not clearly support this viewpoint. *P. leucopus* individuals, blinded by the severing of their optic nerves, homed as well as controls, perhaps because the experimental animals compensated by using some remaining sensory modality which would usually be of lower priority for usage (Parsons and Terman, 1976). A similar priority of cue use concept has been suggested for pigeons (Keeton, 1974). It is also possible, but probably unlikely, that Parsons and Terman's deermice were not really totally blind when they were released for homing. There was no internal anatomical or histological postoperative check on the effectiveness of the blinding operation, but the deermice did have opaque and wrinkled eyes.

Terrain visual cues, however, did not seem to be used by *Apodemus sylvaticus*, for its returns were not enhanced in areas with diversified terrain cues (Bovet, 1965a). This does not necessarily imply that this species might not use terrain visual cues. Saint Girons and Durup (1974) found smaller home ranges in heterogenous areas, perhaps because of more numerous shelter and food sites. Thus, in the Bovet (1965a) study, his subjects may have been displaced outside an area of prior familiarization.

Evidence for visual use of celestial cues is weak and generally negative. Clear moonlit nights inhibited movements of *Peromyscus gossypinus* on a golf course (Griffo, 1961). *Peromyscus* can apparently home successfully under overcast skies (Rawson, 1956). Laboratory studies on *Peromyscus'* possible use of the moon for orienting have yielded contradictory conclusions. *P. leucopus* least preferred a luminous cue to orient by, perhaps even tending to avoid it (Joslin, 1971). Kavanau (1968), however, has concluded that *Peromyscus* does use the moon, for its running wheel activities were frequently oriented to a luminous object. How closely either study's luminous object approximated the moon is open to question. Celestial cue use is not indicated for *Reithrodontomys megalotis* or *Microtus californicus*. Neither one showed initial orientation on release in the field from a circular platform (Fisler, 1967). However, sun compass use has been demonstrated for *Apodemus agrarius* (Lüters and Birukow, 1963). They trained this species to

orient to a given compass direction at a certain time of day by using
an artificial sun; when tested at various other times of day, it exhibited
the appropriate shift in compass direction relative to the time of day.

2. *Olfactory cues*

There is apparently no direct evidence for the use of olfactory cues
by rodents in piloting over long distances. Bovet (1968) considered
and discounted the possibility of his deermice following some gradient
of social odors to the home site. After individually releasing *Peromyscus
maniculatus*, he found a complete absence of other *Peromyscus* activity
in the snow in all directions for at least 30 m around the release point.
However, short distance orientation (of 6 m or less) to the smell of
buried seeds was exhibited by *P. maniculatus*. The percentage of
buried seeds discovered by these mice did not differ significantly
between dim light and total darkness, thus excluding the possible use
of visual cues to explain the results (Howard *et al.*, 1968). Another
Peromyscus species, *P. leucopus*, discriminated fairly well between oat-
baited and empty track shelters (Sheppe, 1965b), presumably on the
basis of odor. Olfaction is not clearly used by *P. leucopus* in homing,
however. Individuals rendered anosmic with 5% zinc sulfate injections
homed as well as did saline-injected controls, perhaps because they
switched to utilization of cues of some other sensory modality (Parsons
and Terman, 1976). Unfortunately, no deermice were rendered both
anosmic and blind. Although there was no anatomical or histological
postoperative check on the effectiveness of the zinc sulfate treatment,
the subjects probably were indeed rendered anosmic. Preoperative
individuals strongly chose whichever side of an olfactometer had no
odor over the side with para-di-chloro-benzene (mothballs); after the
operation, individual deermice showed no significant choice.

Saint Girons and Durup (1974) have suggested that *Clethrionomys
glareolus* uses olfactory cues in homing. They maintained that this
species had numerous but delayed returns from one direction because
poor cover forced a detour from a potential straight line homing route,
and that subsequent direct homeward orientation was based on olfactory
cues (the wind blew in a constant direction during their study). However, there was no evidence to support the assertion that their voles
initially took a certain detour route, and their tabular evidence does
not clearly suggest numerous returns relative to other directions.

3. *Auditory cues*

There is scant and weak evidence for use of auditory cues in piloting.
The clanging of feed trough doors may have been used for piloting
by three *Peromyscus gossypinus* (Griffo, 1961). This supports the

suggestion by Bovet (1968) that direct perception of home site stimuli is sometimes used in piloting. *Peromyscus* conceivably could also orient by echolocating, for it can hear well into the ultrasonic range, up to 100 kHz (Dice and Barto, 1952). At least for rats there is some direct evidence for echolocation. In a maze, blinded rats were able to discriminate between an open alley and one blocked by a barrier. Rats with their hearing impaired performed at about chance level (Riley and Rosenzweig, 1957).

Saint Girons and Durup (1974) suggested that *Clethrionomys glareolus* piloted to auditory cues. Their voles, released near a road and with the home site also near the road, returned home quite rapidly. However, the voles' homing performance was also good for certain home sites and displacement sites far from the road. Furthermore, since traffic presumably traveled in both directions on the road, possible use of an auditory gradient would seem to be difficult.

4. *Tactile cues*

Apparently there is no evidence for use of tactile cues in rodent homing. However, some rodents do seem to use tactile cues at least for short distance orientation. *Peromyscus* in total darkness still maintained an oriented direction in running wheels (Kavanau, 1968), and still chose the artificial tree or artificial grass habitat appropriate for the subspecies (Harris, 1952). *Peromyscus* may also have used tactile cues in the development of strong position habits during a brightness discrimination task (Moody, 1929). Wild rats in a large enclosure may have relied heavily on tactile cues for orienting (Calhoun, 1962).

5. *Kinesthetic cues*

There is also apparently no positive evidence for use of kinesthetic cues in rodent homing. Robinson and Falls (1965) asserted that voles taken by indirect routes prior to release homed as well as controls. However, they did not provide data to support the implied non-usage of kinesthetic cues in homing.

Some rodents do seem to use kinesthetic cues at least for short distance orientation. *Peromyscus* may have used kinesthetic cues (as well as tactile cues) in the Moody (1929) discrimination study. Wild rats may have used kinesthetic cues to a minor extent in orienting in a large enclosure (Calhoun, 1962).

C. NAVIGATION AND INITIAL ORIENTATION

Navigation of some sort is another possible mechanism for rodents to use while homing. Navigation would include the ability to sense one's exact geographical position on displacement, know the "map" coordinates of the home site, and then move along a rather direct course between these two known points. Navigation could be based on proprioceptor input (inertial navigation), or on information acquired from celestial bodies' locations, or from magnetic or other geophysical gradients. However, neither laboratory nor field studies provide compelling evidence for rodent navigation.

The laboratory experiments on initial orientation by Lindenlaub (1955, 1960) and Bovet (1960) at best provide only weak evidence for navigational ability for *Microtus arvalis* and *Apodemus sylvaticus*. These rodents, captured in their home ranges, were released in the laboratory at the center of labyrinth mazes with 24 or 18 exits along the periphery. Displacement distances from the home site varied between 70 m and 3·7 km. The rodents' initial homing orientation seemed rather weak, for statistical significance was obtained in only 6 of the 54 experimental series (Bovet, 1960), or by pooling the data of several experimental series (Lindenlaub, 1955). Furthermore, Lindenlaub's results may be explained by the rodents' possible orientation to some olfactory, auditory, or other cue associated with the nearby sea, rather than to navigational ability (Bovet, 1960). Although Bovet's experiments were not near the sea, perhaps a similar type of explanation could apply to his results. It is questionable how valuable a navigational mechanism would be that is affected adversely by so many variables as Bovet's Table 2 suggests. Odor traces left in his apparatus may also have influenced results.

Bovet (1965b) was able to some extent to train *Apodemus sylvaticus* to choose various directions in a maze. This rather weakly demonstrated ability might not be based on navigational mechanisms, however. Placement of elements in the experimental room was heterogeneous enough to have perhaps allowed for echolocation.

No initial homing orientation was present for *Microtus californicus* or *Reithrodontomys megalotis* released from the center of a circular platform placed in a grassland (Fisler, 1967). According to Bovet (1971), Fisler's negative results could be due to habituation (Fisler's subjects received ten trials in 24 hours; Bovet's and Lindenlaub's, one); to seasonal effects; and to the presence of distracting habitat and celestial cues, which could elicit cover-seeking tendencies instead of initial homeward orienting.

In a winter field release study, initial orientation in the homeward direction (defined as ±90° from the home site direction) has been exhibited by *Peromyscus maniculatus* when released on snow from 100–500 m displacement distances (Bovet, 1971). Average angular errors from the homeward direction were no greater than 37° for 20 m to 80 m distances from the release site (Bovet, 1968). However, another field release study with the same species (Bovet, 1972) showed an absence of homeward-directed initial orientation movements. Bovet attributed these negative results to summer release and the use of a light to observe the deermice. Such differences might be due, however, to familiar cues being more readily perceivable in winter.

Peromyscus maniculatus perhaps exhibited significant homeward orientation, $\pm 90°$, in another field release study (Rawson, 1966), but no data were provided for evaluation. *Clethrionomys glareolus* also exhibited an initial directional tendency toward its home (Durup *et al.*, 1973). Most field release studies, however, indicate a lack of initial orientation in the homeward direction for various rodents. Such negative results have been obtained for *Peromyscus maniculatus* (Murie, 1963), *P. polionotus* (Gentry, 1964), *Microtus pennsylvanicus* (Robinson and Falls, 1965), and *Apodemus sylvaticus* (Lehmann, 1956; Durup *et al.*, 1973).

The positive results obtained by Bovet (1968) may have been due to special conditions. *Peromyscus maniculatus* individuals were released on an open snow-covered plain, with low temperatures. These conditions, Bovet suggested, heightened the motivation of his deermice and prevented their seeking cover and food near the release site (since these "diverting" cues were absent). This postulated tendency to seek cover when it is available is supported by direct observations of released *P. polionotus* (Gentry, 1964).

Bovet's winter homing study of *Peromyscus maniculatus* does not necessarily imply navigation, although this mechanism seemed most likely to him (Bovet, 1968). His data do not suggest a mechanism as precise as navigation for the following reasons. The homing success rate was only 70%. Furthermore, observed movements of the deermice involved large distance zig-zagging, some extensive retracing of routes, and a traversing of *at least* twice the straight line distance home (only part of the total homeward movement could be followed).

Instead, it seems more likely that Bovet's deermice were orienting to partially familiar cues in the area of their home granary. Although these granary deermice were sedentary during the winter, they became less sedentary toward the end of March (Bovet, 1970). Familiarity with the surrounding areas could have been gained during the previous spring and summer. Such familiarity would necessitate a good memory; this has been demonstrated for *P. gossypinus* by Griffo (1961). Indeed, Lehmann (1956) has suggested that his mice homed better in winter because of greater prior acquired familiarity with surrounding areas.

Bovet's displacement distances certainly do not preclude such familiarity. His mice were successively released at 100, 200, 300, 400, and 500 m. Various data reviewed by Stickel (1968) support the possibility of such familiarity for grassland sub-species of *Peromyscus maniculatus*. For instance, the deermice had home ranges with half the observed length (or a radius) of 37–117 m. Thus, a deermouse displaced at 100 m could already be familiar with the area; after its release at 200 m it could orient at some point to familiar cues in the 100 m area, concurrently gaining some experience with the cues in the 200 m area. Likewise, this process could occur for the other increasing displacement distances. Also, juvenile dispersal distances were frequently between 180 and 340 m. Lastly, adult movements of at least 340 m outside the home range have occurred.

Certainly, sufficient cues were present for Bovet's deermice to be familiar with and orient to. These included the granaries themselves, wheat stubble, straw piles, fences, small hill crests, electric lights in farms, and the glow of lights from a nearby city. Kinesthetic cues during transport to the release site are another possibility.

VI. CONCLUSIONS

In spite of numerous homing studies, there is still mostly only indirect evidence for the perceptual mechanisms and scant evidence for the cues that may be used in homing. Non-random movements play a major rôle in homing, but there is yet no convincing evidence that this non-randomness is based on navigation. Familiarity with a large area and use of terrain landmarks for orienting (piloting) probably account for most homing. However, the familiarization process itself has not been experimentally investigated. Instead, the familiarization hypothesis has been supported by sex and age differences in homing, practice effects, and other indirect types of evidence.

Several changes in the methods prevalently used could enhance our knowledge of rodent homing. A method of measuring homing with a time unit more accurate than the usual 12 or 24 hour time unit used in live trap recapture studies would be desirable. More frequent use of direct observation of homing rodents, using techniques similar to those of Griffo (1961) and Bovet (1968), would be helpful. Radiotelemetric techniques may be worth further exploration, in spite of their cost and technical problems. Most importantly, rather than the current *post hoc* discussions, progress in homing research could occur by the designing of experiments specifically to test hypotheses about mechanisms or cues by using the strong inference method.

I propose the following experimental paradigm for a strong inference test of homing mechanisms. This paradigm is similar to one used with bank swallows (Sargent, 1962). Various experimental groups of island rodents would be given varying degrees of exposure to mainland areas. (It is unlikely that the island rodents would have had prior familiarity with the mainland.) Group A would consist of caged individuals moved every several days to points successively farther from the island, with the island perceivable from the closest point. Group B individuals would be given several days' exposure each, at only the nearer points. Group C individuals would be given no mainland experience prior to release.

Live traps could be placed in a circle at an appropriate distance from the release site, the center. These live traps would have a timing device to indicate when the trap was entered. Or, radiotelemetry could be used.

If homing is based entirely on random search, all three groups would be observed with equal frequencies and with similar homing speeds in all directions. If navigation is used, all three groups would be observed with equal frequencies and with similar homing speeds, but most individuals would be observed in an arc facing the island. If area familiarization (based on visual, auditory, or olfactory cues) is involved, Group A should exhibit the best homeward direction performances, Group B should be intermediate, and Group C should be poorest.

If the latter result is obtained, then the relative rôles of terrain *versus* horizon and celestial cues could be ascertained by providing to three groups successively distant exposures on the mainland before release. One group could be housed in cages with an opaque top and upper sides, allowing a view only of terrain cues. Another group could be housed in cages with an opaque lower portion, allowing a view only of horizon and celestial cues. The third group could have an unlimited view. Conclusions would be based on frequencies of observations in various directions and on homing speeds, using logic similar to that discussed above.

Cues and sensory modalities used in homing should be investigated using both laboratory and field manipulative techniques. Although laboratory studies would necessarily involve a study of short distance orientation, it is entirely possible that the sensory modalities and cues used for home range orientation may be the same as those used for homing. Cue preferences within a sensory modality and sensory modality preferences could be investigated using a variety of techniques including simultaneously opposing two previously paired cues for orienting (Joslin, 1971), or by temporarily eliminating a sensory modality (such as plugging the ears or testing in total darkness). Surgical removal of a sense organ probably would not be a good experimental procedure here, because it might result in more than just a sensory deficit. It

D

might also cause degeneration of tracts further up the sensory pathway, including motivational areas.

By using similar methods, controlled field manipulative techniques could be used to compare with the results of laboratory studies. The cues should be easily manipulable by the experimenter. As with the laboratory studies, the field manipulative studies may require some training of the rodents and the use of cues as similar as possible to natural objects such as trees, logs, and forbs. Such manipulative studies are to be preferred over naturalistic field studies whose data could be subject to diverse interpretation. Further understanding of rodent homing is most likely to occur with increased use of such manipulative techniques.

ACKNOWLEDGEMENTS

I wish to thank M. Balaban, R. Baker, R. Raisler, and J. King for their reading of the earliest stages of this manuscript. (This review is a modified version of one written in partial fulfillment of the requirements for the Ph.D. degree, Michigan State University, East Lansing.) Special thanks go to J. King for his comments on a subsequent state of this review. My wife Dottie has helped by giving suggestions pertaining to grammatical and stylistic details.

REFERENCES

Adler, H. E. (Ed.) (1971). Orientation: sensory basis. *Ann. N.Y. Acad. Sci.* **188**, 408 pp.

Blair, W. F. (1951). Population structure, social behavior and environmental relations in a natural population of the beach mouse (*Peromyscus polionotus leucocephalus*). *Contr. Lab. vertebr. Biol. Univ. Mich.* **48**, 1–47.

Bovet, J. (1960). Experimentelle Untersuchungen über das Himfindevermögen von Mäusen. *Z. Tierpsychol.* **17**, 728–755.

Bovet, J. (1962). Influence d'un effet directionnel sur le retour au gîte des mulots fauve et sylvestre (*Apodemus flavicollis* Melch et *A. sylvaticus* L.) et du campagnol roux (*Clethrionomys glareolus* Schr.) (Mammalia, Rodentia). *Z. Tierpsychol.* **19**, 472–488.

Bovet, J. (1965a). Note sur le retour au gîte du Mulot en Camargue. *Z. Tierpsychol.* **22**, 1963–1966.

Bovet, J. (1965b). Ein Versuch, wilde Mäuse unter ausschluss optischer, akustischer und osmicher Merkmale auf Himmelsrichtungen zu dressieren. *Z. Tierpsychol.* **22**, 839–859.

Bovet, J. (1968). Trails of deermice (*Peromyscus maniculatus*) traveling on the snow while homing. *J. Mammal.* **49**, 713-725.

Bovet, J. (1970). Sedentariness of *Peromyscus maniculatus* in Alberta granaries. *J. Mammal.* **51**, 632–634.

Bovet, J. (1971). Initial orientation of deer mice (*Peromyscus maniculatus*) released on snow in homing experiments. *Z. Tierpsychol.* **28**, 211–216.

Bovet, J. (1972). Displacement distance and quality of orientation in a homing experiment with deermice (*Peromyscus maniculatus*). *Can. J. Zool.* **50**, 845–853.

Brant, D. H. and Kavanau, J. L. (1964). "Unrewarded" exploration and learning of complex mazes by wild and domestic mice. *Nature (Lond.)* **204**, 267–269.

Brant, D. H. and Kavanau, J. L. (1965). Exploration and movement patterns of the canyon mouse *Peromyscus crinitus* in an extensive laboratory enclosure. *Ecology* **46**, 452–461.

Burt, W. H. (1940). Territorial behavior and populations of some small mammals in southern Michigan. *Misc. Publs. Mus. Zool. Univ. Mich.* **45**, 1–58.

Calhoun, J. B. (1962). "The Ecology and Sociology of the Norway Rat." U.S.D.A.H.E.W., Public Health Serv., 288 pp.

Chitty, D. (1937). A ringing technique for small mammals. *J. Anim. Ecol.* **6**, 36–53.

Davis, R. (1966). Homing performance and homing ability in bats. *Ecol. Monogr.* **36**, 201–237.

De Busk, J. and Kennerly, T. E., Jr. (1975). Homing in the cotton rat, *Sigmodon hispidus*. *Am. Midl. Nat.* **93**, 149–157.

Dice, L. R. and Barto, E. (1952). Ability of mice of the genus *Peromyscus* to hear ultrasonic sounds. *Science, N.Y.* **116**, 110–111.

Durup, M., Saint Girons, M. C., Fabrigoule, C. and Durup, H. (1973). Quelques données sur les modalités du retour au gîte chez le mulot, *Apodemus sylvaticus*, et le campagnol roussatre *Clethrionomys glareolus*. *Mammalia* **37**, 34–55.

Fisler, G. F. (1962). Homing in the California vole, *Microtus californicus*. *Am. Midl. Nat.* **68**, 357–368.

Fisler, G. F. (1966). Homing in the western harvest mouse, *Reithrodontomys megalotis*. *J. Mammal.* **47**, 53–58.

Fisler, G. F. (1967). An experimental analysis of orientation to the homesite in two rodent species. *Can. J. Zool.* **45**, 261–269.

Furrer, R. K. (1973). Homing of *Peromyscus maniculatus* in the channelled scablands of east-central Washington. *J. Mammal.* **54**, 466–482.

Galler, S. R., Schmidt-Koenig, K., Jacobs, G. J. and Belleville, R. E. (Eds.) (1972). "Animal Orientation and Navigation." *NASA SP*-262.

Gentry, J. B. (1964). Homing in the old-field mouse. *J. Mammal.* **45**, 276–283.

Grant, P. R. (1971). Experimental studies of competitive interaction in a two-species system. III. *Microtus* and *Peromyscus* species in enclosures. *J. Anim. Ecol.* **40**, 323-350.

Griffin, D. R. and Hock, R. J. (1949). Airplane observations of homing birds. *Ecology* **30**, 176–198.

Griffo, J. V., Jr. (1961). A study of homing in the cotton mouse, *Peromyscus gossypinus*. *Am. Midl. Nat.* **65**, 257–289.

Hamilton, W. J., Jr. (1937). Activity and home range of the field mouse, *Microtus p. pennsylvanicus* (ord.). *Ecology* **18**, 255–263.

Harris, V. T. (1952). An experimental study of habitat selection by prairie and forest races of the deermouse, *Peromyscus maniculatus*. *Contr. Lab. vertebr. Biol. Univ. Mich.* **56**, 1–53.

Howard, W. E., Marsh, R. E. and Cole, R. E. (1968). Food detection by deermice using olfactory rather than visual cues. *Anim. Behav.* **16**, 13–17.

Joslin, J. K. (1971). "Visual cues used in orientation by whitefooted mice, *Peromyscus leucopus*: a laboratory study." Ph.D. Thesis, Michigan State Univ., 134 pp.

Kavanau, J. L. (1968). Activity and orientational responses of white-footed mice to light. *Nature (Lond.)* **218**, 245-252.

Keeton, W. T. (1974). The orientational and navigational basis of homing in birds. *In* "Advances in the Study of Behavior", vol. 5 (Eds. D. S. Lehrman, J. S. Rosenblatt, R. A. Hinde and E. Shaw). Academic Press, New York, pp. 48–132.

King, J. A. (1968). Psychology. *In* "Biology of *Peromyscus* (Rodentia)" (Ed. J. A. King). Spec. Publ. No. 2, Am. Soc. of Mammalogists, pp. 496–542.

King, J. A. and Vestal, B. M. (1974). Visual acuity of *Peromyscus*. *J. Mammal.* **55**, 238–243.

Lay, D. W. and Baker, R. H. (1938). Notes on the home range and ecology of the Attwater wood rat. *J. Mammal.* **19**, 418–423.

Layne, J. N. (1954). The biology of the red squirrel, *Tamiasciurus hudsonicus loguax* (Bangs) in New York. *Ecol. Monogr.* **24**, 227–267.

Layne, J. N. (1957). Homing behavior of chipmunks in central New York. *J. Mammal.* **38**, 519–520.

Lehmann, E. von (1956). Heimfindeversuche mit kleinen Nagern. *Z. Tierpsychol.* **13**, 485-491.

Lindenlaub, E. (1955). Über das Heimfindevermögen von Säugetieren. II: Versuche an Mäusen. *Z. Tierpsychol.* **12**, 452–458.

Lindenlaub, E. (1960). Neue Befunde über die Anfangsorientierung von Mäusen. *Z. Tierpsychol.* **17**, 555–578.

Lüters, W. and Birukow, G. (1963). Sonnenkompassorientierung der Brandmaus (*Apodemus agrarius* Pall). *Naturwissenschaften* **50**, 737–738.

McCabe, T. T. and Blanchard, B. D. (1950). "Three Species of *Peromyscus*". Rood Associates, Santa Barbara, 136 pp.

Mallach, N. (1972). Translokationsversuche mit Bisamratten (*Ondatra zibethica* L.). *Anz. Schädlingsk Planzenschutz* **45**, 40–44.

Metzgar, L. H. (1967). An experimental comparison of screech owl predation on resident and transient white-footed mice (*Peromyscus leucopus*). *J. Mammal.* **48**, 387–391.

Metzgar, L. H. (1973) Home range shape and activity in *Peromyscus leucopus*. *J. Mammal.* **54**, 383–390.

Moody, P. A. (1929). Brightness vision in the deer-mouse, *Peromyscus maniculatus gracilis*. *Exp. Zool.* **52**, 367–405.

Morris, R. D. (1967). A note on the homing behavior of *Peromyscus maniculatus osgoodi*. *Can. Fld Nat.* **81**, 225–226.

Murie, M. (1963) Homing and orientation of deermice. *J. Mammal.* **44**, 338-349.

Murie, O. J. and Murie, A. (1931). Travels of *Peromyscus*. *J. Mammal.* **12**, 200–209.

Orr, R. (1970). "Animals in Migration," Macmillan, New York, 291 pp.

Parsons, L. and Terman, C. R. (1976). The influence of vision and olfaction on the homing ability of the white-footed mouse, *Peromyscus leucopus noveboracensis*. Presented at Va. Acad. Sci. Meetings, May 1976.

Rawson, K. S. (1956). "Homing activity and endogenous activity rhythms." Ph.D. Thesis, Harvard Univ., 111 pp.

Rawson, K. S. (1966). Goal directed orientation in the homing behavior of mice (genus *Peromyscus*). *Bull. ecol. Soc. Am.* **47**, 199.

Rawson, K. S. and Hartline, P. H. (1964). Telemetry of homing behavior by the deermouse, *Peromyscus*. *Science, N.Y.* **146**, 1596–1598.

Riley, D. A. and Rosenzweig, M. R. (1957). Echolocation in rats. *J. comp. physiol. Psychol.* **50**, 323–328.

Robinson, W. L. and Falls, J. B. (1965). A study of homing of meadow mice. *Am. Midl. Nat.* **73**, 188–224.

Saint Girons, M. C. and Durup, M. (1974). Retour au gîte chez le mulot, *Apodemus sylvaticus*, et le campagnol roussatre, *Clethrionomys glareolus*: facteurs écologique, apprentissage et mémoire. *Mammalia* **38**, 389–409.

Sargent, T. D. (1962). A study of homing in the bank swallow (*Riparia riparia*). *Auk* **79**, 234–246.

Savidge, I. R. (1973). A stream as a barrier to homing in *Peromyscus leucopus*. *J. Mammal.* **54**, 982–984.

Schleidt, W. (1951). Orientierende Versuche über die Heimkehrfähigkeit der Rotelmaus (*Evotomys glareolus ruttneri* Wottst.). *Z. Tierpsychol.* **8**, 132–137.

Schmidt-Koenig, K. (1965). Current problems in bird orientation. *In* "Advances in the Study of Behavior", vol. 1 (Eds. D. S. Lehrman, R. A. Hinde and E. Shaw). Academic Press, New York, pp. 217–278.

Sheppe, W. (1965a). Dispersal by swimming in *Peromyscus leucopus*. *J. Mammal.* **46**, 336–337.

Sheppe, W. (1965b) Characteristics and use of *Peromyscus* tracking data. *Ecology* **46**, 630–634.

Sheppe, W. (1966). Determinants of home range in the deer mouse, *Peromyscus leucopus*. *Proc. Calif. Acad. Sci.* **34**, 377–418.

Smith, D. A. and Speller, S. W. (1970). The distribution and behavior of *Peromyscus maniculatus gracilis* and *Peromyscus leucopus noveboracensis* (Rodentia: Cricetidae) in a southeastern Ontario woodlot. *Can. J. Zool.* **48**, 1187–1199.

Stickel, L. F. (1949). An experiment on *Peromyscus* homing. *Am. Midl. Nat.* **41**, 659–664.

Stickel, L. F. (1968). Home range and travels. *In* "Biology of *Peromyscus* (Rodentia)" (Ed. J. A. King). Spec. Publ. No. 2, Am. Soc. of Mammal., pp. 373–411.

Stickel, L. F. and Warbach, O. (1960). Small-mammal populations of a Maryland woodlot, 1949–1954. *Ecology* **41**, 269–286.

Storm, R. M. (Ed.) (1967). "Animal Orientation and Navigation." Oregon State Univ. Press, Corvallis, Oregon.

Terman, C. R. (1962). Spatial and homing consequences of the introduction of aliens into semi-natural populations of prairie deer-mice. *Ecology* **43**, 216–223.

Terman, C. R. (1968). Population dynamics. *In* "Biology of *Peromyscus* (Rodentia)" (Ed. J. A. King). Spec. Publ. No. 2, Am. Soc. of Mammal., pp. 412–450.

Vestal, B. M. (1973). Ontogeny of visual acuity in two species of deermice (*Peromyscus*). *Anim. Behav.* **21**, 711–719.

Secondary Production in Inland Waters[1]

THOMAS F. WATERS

*Department of Entomology, Fisheries and Wildlife,
University of Minnesota,
St Paul, Minnesota 55108*

I. INTRODUCTION

For the individual organism, production is the growth of its own body. From the standpoint of the functioning of an ecosystem, production is the means by which energy is made available for transmission from one trophic level to the next. For a consumer, including humans, it is the production of lower trophic levels upon which it depends for its trophic sustenance.

Small wonder then that both scientists and resource managers have been intensely concerned with the processes of production and, more

[1] Paper No. 9959, Scientific Journal Series, Minnesota Agricultural Experiment Station, St Paul, Minnesota 55108.

91

recently, with measurement of the rates of production. In his original textbook on limnology, Welch (1935) defined his discipline ". . . as that branch of science which deals with biological productivity of inland waters . . .".

Furthermore, the measurement of production rates of all trophic levels has taken on great significance with the increased concern for the quantification of ecosystem dynamics, since production is one of the major pathways of energy flow. Additionally, secondary production is of vital interest to natural resource administrators responsible for the management of wild populations utilized for both food and recreation, and in inland waters, particularly for vertebrate fisheries.

Whereas much knowledge has been gained world-wide on primary production, in both freshwater and marine environments, accumulation of data on secondary production until recently has lagged far behind. Unlike methods for estimating primary production rates (photosynthesis) which utilize the release or uptake of single specific gases or ions in the aquatic medium, no such method for estimating secondary production exists; even if it did, such a method would not usually be valid, because secondary producers include more than one trophic level, and production rates for different trophic levels are not additive.

Several reasons for this lag are apparent. One is the difficulty in applying theoretical models to a group of organisms with diverse and complex life histories; in any given fauna, the life history of all member species may not even be determinable because of the lack of taxonomic capabilities for identifying aquatic larval stages. Secondly, all methods require the collection of large numbers of samples from the population because of great variation in distribution, resulting in high sampling error; there is the additional problem of possible systematic error in sampling which is difficult to deal with or to measure, or frequently even to detect. Finally, the sorting and analysis of the samples—the "bug-picking" job—usually is so tedious and expensive of research resources that this problem alone may discourage or preclude an important project. Consequently, most basic research on energy flow through secondary producers (defined here as all heterotrophic levels) has been involved with methodological development. And a considerable portion of this effort has been concerned with the problem of alleviating the sample-sorting chore.

In his definitive work on stream ecology, which carries a review of literature through 1966, Hynes (1970) lists only four attempts at the direct estimation of benthic invertebrate production in streams. He concluded rightfully that at that time this area of investigation was in its infancy.

The most recent review of production in the aquatic ecosystem as a whole is that of K. H. Mann (1969), who dealt with both processes and methods, with all trophic levels, and in both freshwater and marine environments, both lakes and streams, up to approximately the same date. And while work on secondary production generally began somewhat earlier in lakes than it did in streams, it is clear that aside from pioneer works dealing with conceptual struggles, most work on the subject has been done in approximately the last decade, a point clearly brought out by the perusal of literature dates in the present review.

Yet within this last decade, great strides forward have been made— on conceptual understanding of production biology, on method development, and in the accumulation of reliable data. Most recently, secondary production rates have begun to be used in the evaluation of environmental changes.

It seems appropriate at this time then to deal with a selected portion of aquatic production biology—secondary production in inland waters —with emphasis, necessarily, on the advances made roughly in the last ten years. Even among the secondary producers, this review is limited; while functionally the term secondary production is understood to include all heterotrophic organisms, this review will deal mainly with three groups—fishes, zoobenthos, and zooplankton. These are the groups which have received most attention, for which methodology has been best worked out, and which are of most immediate concern to man. The writer's bias towards benthos and fish as opposed to zooplankton, probably is also apparent. The production of other secondary producers, including heterotrophic algae, bacteria and fungi, protozoans, amphibians, reptiles, and other aquatic vertebrates such as some birds and mammals more commonly associated with the terrestrial environment, must await further advances and reviews by other specialists. The energy flow through these other groups, of course, may be extremely important.

Much good work involving ecological principles on the production biology of all groups has been done under experimental circumstances, e.g. the extensive pond experiments by Hall et al. (1970). For the most part, the data obtained on production rate under experimental conditions have not been included in the summaries, the primary reason being that the levels of production so obtained are not comparable, with some noted exceptions, to levels observed under natural conditions. These include caged or bottled specimens, aquarium experiments, stocked populations (such as hatchery-reared fish), and organisms in artificial water bodies such as laboratory streams, constructed experimental ponds, or those receiving artificial feeding or fertilization. On the other hand, production data are included when they have been specifically used to evaluate management programmes, such as physical habitat improvement or pollution abatement.

In addition to production data actually published as such, many papers in the literature contain data on population dynamics, life tables, or physiological data from which production rates could, with some assumptions, be computed (e.g. Gehrs and Robertson, 1975). This I have not done; it seemed to me that when the estimation of production rate was not the original objective, the methods used frequently involved so much personal interpretation on the part of the researcher that it would have been too risky to make the calculations without a high possibility of serious systematic error.

A review of production for any organism group or in any environment would be remiss if it did not acknowledge the enormous stimulation provided to the study of production biology by the International Biological Programme. Indeed, the advances made in the last decade have been largely the result of the emphasis of IBP on production biology. Accomplishments have included many meetings, conferences and symposia held throughout the world, frequently bringing together scientists with common ecological goals from widely divergent political areas and language origins. Proceedings of most of these have been published, including those edited by Goldman (1966), Petrusewicz (1967), Gerking (1967), Petrusewicz and Ryszkowksi (1969/1970), Fuller and Kevan (1970), Winberg (1971), Edwards and Garrod (1972), Kajak and Hillbricht-Ilkowska (1972), National Academy of Sciences (1975), Mori and Yamamoto (1975), and Salanki and Ponyi (1975).

Other direct results of the IBP have included a series of valuable "handbooks" on methods applicable to various groups, including those on large terrestrial herbivores (Golley and Buechner, 1968), fishes (Ricker, 1968), primary production (Vollenweider, 1969), terrestrial animals (Petrusewicz and Macfadyen, 1970), aquatic secondary production (Edmondson and Winberg, 1971), marine benthos (Holme and McIntyre, 1971), and microbial production (Sorokin and Kadota, 1972).

Additionally, the IBP has stimulated much other work, directly or indirectly, the published accounts of which have appeared in many regular scientific journals.

Other recent reviews on secondary production include chapters in the books by Russell-Hunter (1970) and Warren (1971), Le Cren (1972) on fish production in freshwaters, and Johnson (1974) on the production of lake benthos. Edmondson (1974) has provided an analysis of some of the limitations and possible violations of assumptions that may lead to serious errors in production estimates, particularly with zooplankton.

II. Terminology

Ecologists have agonized for decades over the terminology of production biology, with respect to basic concepts as well as to specific definitions and the introduction of variants. As a consequence, much care must be used, particularly in the older (pre-World War II) literature, in interpreting published data. The terms "productivity" and "production" have been used with seeming endless definitions, from static quantities present, to fertility, to "potential" production, to yield of fisheries, to productive processes, etc. However, D. J. Crisp (1971) and Batzli (1974) have provided discussions of current concepts and terms.

In this review, only three terms will be defined and employed: (1) standing stock, (2) production and (3) yield.

Standing stock is defined as the amount present at a point in time, expressed best as quantity per spatial unit. Synonyms include standing crop, stock, and biomass. The last term, biomass, is probably used most frequently, particularly since the IBP-recommended symbol for standing stock, B, is taken from it. However, the term biomass has its own specific meaning (the weight of organism matter on earth), and could be easily taken to mean something other than standing stock. Units employed for standing stock include weight (usually either grams or kilograms), either wet or dry weight or ash-free dry weight, carbon, and calories, per m^2 or ha (see Table I). Expressions for zooplankton are sometimes given in units of m^3, because it is in this spatial unit that raw data are acquired, but unless the depth, or better, thickness of the productive stratum, is known, such expressions are impossible to relate to other trophic levels or to data from other environments. In fisheries work, the most common unit appears to be kg/ha (wet weight). In some cases, standing stock is expressed for the entire lake, stream section, etc., without reference to a spatial unit, but this has obvious drawbacks from the standpoint of comparison with other water bodies or the analysis of density. Standing stock expressed in terms of numbers of organisms has the obvious disadvantage of difficulty in handling variation in individual mean weight, and the lack of comparability with other populations; it is probably useful only in applications where numbers are of principal concern, e.g. number of salmon smolts. Expressions such as "monthly standing stock" are confusing and inappropriate, because these incorrectly imply a *rate*; probably, mean monthly standing stock is the intention of such term, in which case it should be stated explicitly as a mean. The most commonly used units appear to be g/m^2 (dry), kg/ha (wet or dry), $kcal/m^2$, and units of carbon and ash-free dry weight. Obviously, units used should be explicitly stated (including wet, dry or ash-free dry weight) in all abstracts, summaries, tables and figures.

Production is defined as the rate of tissue elaboration, regardless whether it survives to the end of a given period of time, in the classic sense of Ivlev (1945) and Clarke (1946). It is expressed in units of quantity/spatial unit/unit time. The units employed are the same as for standing stock with the addition of the time unit; thus, *production* always implies a rate. The most commonly used units appear to be kg/ha/yr (wet weight) for fish, kg/ha/yr (dry weight) for zoobenthos, and $g/m^2/yr$ (dry weight) for zooplankton. Furthermore, with zooplankton, production is frequently expressed on a daily basis. With fish, wet weight is the more common. An obvious problem comes in when dealing with estimates of production over a period of time that is long relative to generation time or life span (e.g. annual production for multivoltine zooplankters), namely, that materials may be recycled during the time for which production is estimated. Under these circumstances, it is possible that, were there to be harvest and removal within the time period, production might not be as high as estimated. Consequently, there is an increasing use of the caloric unit, e.g. kcal; the older expression, Cal. (= kcal), should be avoided because of typographical confusion with cal (= kcal/1000). Obviously, the flow of energy through the trophic levels and various other pathways in an ecosystem is much more readily perceived if all rates are expressed in the same caloric units.

There is a tendency among (lake) limnologists to equate "productivity" with primary production only; consequently, many titles of papers and discussion within papers using the terms "productivity" frequently have no direct concern with other than the primary producers. In that case, other units, such as O_2 released or CO_2 taken up, are also employed, but these need not concern us in secondary production. Primary production additionally must be identified as either net or gross, and sometimes these prefixes are stated in secondary production as well; when they have been used, gross production includes the energy of respiration while net production does not. The usual convention in secondary production is, according to Odum (1959), simply to use the term *production* as not including respiratory energy. Secondary production is analogous to net primary production, and gross primary production is analogous to assimilation (production plus respiration) of secondary producers. Obviously, researchers in secondary production have an easier time of it. The same comments made for standing stock above with respect to the need for explicit statements of units apply equally to production.

Yield is defined as the rate of transfer from one population or trophic level to the next higher, or predating, group. Thus, a group "yields" a portion of its production to its predator group. The rate of ingestion by the predator group (higher trophic level) may be equal to the yield of the prey group, but frequently it is not; the predator group may be ingesting food from outside the ecosystem in addition to the yield of the lower trophic level. For example, a carnivorous fish population may be feeding on terrestrially-derived insects as well as the yield from the bottom fauna. The "production" of a commercial fishery is actually "yield", as defined above, i.e. the fishermen's catch. Yield is expressed in the same units as production—quantity/spatial unit/unit time.

III. Methods for Estimating Production Rate

A. removal-summation methods

One of the earliest published accounts of an estimate of secondary production, actually conducted in marine waters, was that of Boysen-Jensen (1919) who estimated the annual production of several benthic invertebrates in a Danish fjord. Credit for the introduction of a conceptually correct method is generally assigned to him. He utilized the idea that what was produced eventually died or was otherwise removed, and the estimation of total mortality of a year class or cohort of an animal was therefore equivalent to an estimate of production for that cohort.

We may term procedures that employ this principle *removal-summation* methods, and these may be grouped into two types:

1. *Iteration of apparent losses*

When a cohort is observed throughout its life span, from hatching of many small instars or fry to the disappearance of the last old survivor, a continuous reduction in numbers will be apparent through time. If an accurate assessment of these losses can be taken and recorded through a series of samples that provides standing stock data, together with an assignment of mean weight to the losses at the time of loss, they may be summed over the entire life span to provide a total estimate of removal (elimination, in the sense of Winberg, 1971), which is then taken as the equivalent of total production of the cohort.

If the species is univoltine, the production so estimated is annual production; if it is multivoltine, it must be added to that of other cohorts produced in the same year to equal annual production. If it is hemivoltine, and there are overlapping cohorts, it is considerably more difficult to ascertain annual production, for the removal during a given year which is part of the cohort life span is not necessarily equal to the production for the same yearly period of time; it is only the total removal of the cohort that is equal to its total production. If such hemivoltine species produce a new generation beginning each year, and if populations are stable year to year, then annual production for any one year would be equal to one cohort production, being the sum of the end of one cohort and the beginning of a second. Frequently, it is necessary to add mortality to changes in standing stock algebraically, since populations are rarely stable year to year.

This removal-summation method was also used by Lundbeck (1926), Sanders (1956) for marine benthos, Anderson and Hooper (1956) in a freshwater lake, Teal (1957) in a temperature zone spring, and Sokolowa (1966). Comita (1972) employed this procedure on planktonic crustaceans, where cohorts could be identified. Waters and Crawford (1973) compared the results from this method to those from other methods on the production of a stream mayfly.

2. Independent estimates of removal

Unlike the previous procedure, which records the gradual disappearance of a population, this variant of a removal-summation process records separate and independent estimates of all the various forms of mortality and/or emigration, such as predation by fish, molting losses, emergence, drift, etc. Borutsky (1939) employed this procedure on the profundal benthos of Lake Beloie by recording emergence, predation by fish (estimated approximately), and "natural" mortality, combining these losses with observed changes in standing stock. Similar procedures on benthos, adapted to different circumstances, were employed by Miller (1941) in a Canadian lake, Odum (1957) in a Florida spring system, Gerking (1962) in an Indiana lake, and Waters (1962, 1966) in a small Minnesota stream. None of these procedures, however, would seem to have a very broadly based applicability, because they all depend heavily on the particular circumstances of the population, the water body, and the techniques available to an investigator under unique conditions.

B. INCREMENT-SUMMATION METHOD

A method similar to the iterative removal-summation procedure in its basic technique is one of summing up the growth increments through the life span of a cohort. In this method, a series of samples is taken periodically through the cohort life, beginning with the population of abundant first hatchlings. From one sample to the next, the mean growth increment is computed as the increase in mean individual weight; this is multiplied by the mean standing stock in numbers for the period to obtain an estimate of production for the period. The cohort production then is the sum of all such products for the entire cohort. Again, annual production is equal to the cohort production for a univoltine species, and to the sum of the several generations for multivoltine species. On the other hand, annual production for hemivoltine species is correctly obtained as the sum of the period increments for the year, irrespective of population stability year to year. Of course, the increment summation should equal removal summation for a cohort.

The origin of this method is most usually assigned to Pechen and Shushkina (1964) and Winberg et al. (1965) and has been variously discussed by Greze (1965), Kajak and Rybak (1966), Kajak (1967), Konstantinov and Nechvalenko (1968), and others. Based on previous works by the author and his associates, Winberg (1971) describes graphical methods by which the increment summation may be accomplished by plotting mean individual weight, specific daily growth increments, and numbers in the cohort population, and from these deriving daily production rates.

C. INSTANTANEOUS GROWTH RATE METHOD

The mathematical foundation of production estimation was provided independently by Ricker (1946) and Allen (1949), resulting in the instantaneous growth rate method. In its simplest form, the method involves the product of instantaneous rate of growth and mean standing stock:

$$P = G\bar{B}$$

where P = production for a given period of time in weight/spatial unit/time period

G = instantaneous growth rate for the time period

$\bar{B}$ = mean standing stock during the time period, in weight/spatial unit

Both Ricker and Allen developed the method for use in fish production studies, and Ricker and Foerster (1948) first used it to compute the production of juvenile sockeye salmon (*Oncorhynchus nerka*) in a Pacific coast lake. However, the method is basically applicable to any population of living organisms, and since its introduction it has also been used widely for invertebrates, and even plants (Mathews and Westlake, 1969) as well as for fish.

The method estimates production for a given interval of time which is appropriate within the animal's life cycle and for which sampling is feasible. During this interval, both growth and mortality must be exponential and remain constant or vary similarly. Consequently, the time period is usually selected short enough that these assumptions can be reasonably expected to be valid; approximately one month has been employed as an interval generally considered satisfactory for both fish and benthic invertebrates, whereas a shorter interval would be preferable for most zooplankters which have a shorter life cycle. If the above assumptions are held to be valid over a longer period of time, the sampling interval may be longer—e.g. one year in the case of fish. Allen (1950) provided an alternate formula to be used when growth is sigmoid, and other formulae for circumstances in which there may be various combinations of growth and mortality of other forms (Allen, 1971). In application, the original form, ($P = G\bar{B}$), has been used almost universally.

Instantaneous growth rate is calculated as the natural logarithm of the ratio of the mean weight at end of the interval to mean weight at the beginning. Thus, a distinct cohort must be identifiable, or the animals must be otherwise aged, from field samples; alternatively, instantaneous growth rate may be determined in laboratory cultures that simulate those conditions that affect growth, such as temperature and food availability. In his well-known work on the metabolism of an entire community of a temperate spring pool, Teal (1957) determined instantaneous growth with laboratory cultures. In the case of fish, aging may be accomplished with annual marks on scales or bones; and with invertebrates aging usually is by following a cohort through size frequency distributions. In some cases, probability paper may be employed to assist in the identification of cohorts (Harding, 1949; Cassie, 1950, 1954), but at least some modal frequency or inflection in the mean size-time curve must be apparent to detect cohorts. Ladle *et al.* (1972) and Burke and Mann (1974) have used probability paper in facilitating growth rate measurements in invertebrates.

Mean standing stock in weight/spatial unit is most usually computed as a simple average of two samples taken at the ends of the time interval, a practice that appears satisfactory when the interval is relatively short. For longer periods (e.g. one year for fish) mean standing stock may be calculated, provided that initial assumptions on the form of growth and mortality are met, by the formula given originally by Ricker and Allen:

$$\bar{B} = B_o \times \frac{(e^{G-Z} - 1)}{G - Z}$$

where $\bar{B}$ = mean standing stock

B_o = standing stock at beginning of interval

G = instantaneous growth rate for interval

Z = instantaneous mortality rate for interval

The term, $\dfrac{e^{G-Z}-1}{G-Z}$, is most easily obtained from the table in Appendix I of Ricker (1975) as follows: If $G - Z$ is positive, enter the table in column 1 (labeled Z) and read the term in column 5; if $G - Z$ is negative enter the table similarly in column 1 (using absolute value of $G - Z$) and read the term in column 4 (see Ricker, 1975, p. 239).

General discussions of the method have been given by Chapman (1967, 1968), Backiel (1967), and Warren (1971).

D. ALLEN CURVE METHOD

Extending the instantaneous growth formula to a graphical representation, Allen (1951) estimated the production of brown trout (*Salmo trutta*) in his classic work on the Horokiwi Stream, New Zealand. The resulting figure is basically a growth-survivorship curve prepared for a given cohort with the number of survivors plotted against mean individual weight (Fig. 1). So constructed, the area under the curve is expressed in units of the axes as cohort production. The standing stock at any point, X, is equal to the shaded area $A'WXY$, the product of number and mean weight. Since the initial point of the curve is theoretically at the number of births and their mean weight, the initial standing stock of the cohort ($A'B'BA$) is more correctly assigned to the previous generation, which produced the eggs. This quantity may be a significant portion of the total area and should be taken into account when precise measures of energy flow are desired. The corresponding production of the cohort, then, is area $ABCD$.

Normally the curve would be prepared for a single distinct cohort, following the numbers and mean weight from time of hatching or birth through the life span of the cohort to the last survivor. However, if year-to-year (or cohort-to-cohort) stability of populations can be assumed, a composite Allen curve can be constructed, using the various portions of the curve representing each age group present; in this case, a single cohort production will be equal to annual production of the mixed-age population. This is actually the procedure that Allen first used with the Horokiwi brown trout, and which has also been used by others (e.g. Cooper and Scherer, 1967).

The method was proposed for invertebrates by Neess and Dugdale (1959), and in recent years it has been widely used for both fish and invertebrates, and also for aquatic plants (Mathews and Westlake,

1969). In a number of cases, it has been tested in conjunction with other methods for invertebrates and compared favorably in most cases (Elliott, 1973; Neveu, 1973; Waters and Crawford, 1973; Décamps and Lafont, 1974; Maitland and Hudspith, 1974; Cushman *et al.*, 1975).

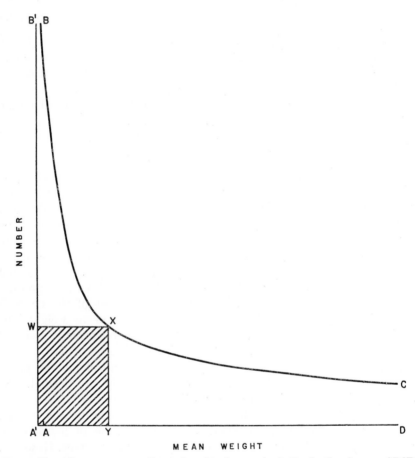

FIG. 1. The Allen curve, growth-survivorship for a cohort. Production is area ABCD; standing stock at point X is area A'WXY. (Reprinted by permission from *The American Naturalist* **103**, 175, 1969.)

The Allen curve method has been given general discussion by Chapman (1967, 1968), Waters (1969), and K. H. Mann (1969, 1971).

E. THE HYNES METHOD

The major practical problem in the estimation of production with any of the above methods is the often preclusive one of cohort identification, in addition to an expenditure of research resources in sample sorting, etc. Towards the solution of these problems, Hynes (1961) introduced an innovative approach. Somewhat similar to the removal-summation methods, his procedure was to sum the losses between successive size groups, rather than the losses between successive times. His objective was to calculate at least an approximation of production rate, for a group of species which could be treated in terms of size units, without the necessity of identifying individual cohorts. In the original paper, Hynes included ten different species, all univoltine, and sampling was conducted over the period of one year; the estimate, therefore, was of annual production.

Conceptual errors in the original were later corrected by Hynes and Coleman (1968) who offered the method as a means to provide "results of the right order of magnitude". Hamilton (1969) made further corrections so that the method is outlined in his paper in the form which now may be generally applied. The essence of it is to sort the organisms from bottom samples into selected size groups (1-mm size groups are convenient for most benthic invertebrates). compute the mean standing stock in numbers over the entire sampling period for each size group, determine the loss in numbers between successive size groups and multiply by mean weight to obtain the loss in weight, multiply by a factor equal to the number of size groups, and sum these products for the estimate of production. It is important that only the sum of the final products be taken as production, rather than considering the individual products for separate size groups; negatives (resulting where there is a "negative loss," i.e. increase in numbers between size groups) should be included in the sum algebraically. Hamilton's Table 2 exemplifies the calculation for Hynes and Coleman's data, in volume units; Table 4 of Waters and Crawford (1973) illustrates the calculation using original weight units.

Several limitations and assumptions must be accounted for in the method, as emphasized by Hamilton:

(1) All species in the group must be of the same voltinism. For univoltine species, the method based on an annual set of samples will estimate annual production; for multivoltine species, the estimate must be multiplied by the number of generations per year to obtain annual production; for those species with life spans of more than one year, the estimate must be divided by the numbers of years in the life span, to obtain annual production.

(2) All species in the group being treated must be capable of growing to the same maximum size. Since the "times loss" factor will be according to the species growing to the largest maximum size, this factor will obviously overestimate the production for species with

less than this maximum size, if they are included. For example, Hamilton applied the corrected method to Hynes's original ten species and found an annual production nearly four times that originally calculated by Hynes. But the ten species were of varying maximum size; when production was calculated separately for each species and summed (by present writer, unpublished), the total was about $\frac{2}{3}$ of that which Hamilton computed. Even with a single species, there may be a difference in maximum size between the sexes, possibly causing an overestimate (Waters and Crawford, 1973).

(3) The growth of the organisms must be assumed to be linear—i.e. the method assumes they will take equal time to grow through each size group. This assumption is perhaps the most difficult to deal with, although Hamilton concluded that the effect is probably not serious.

Since the method in its final form was published, it has stirred both support and criticism. It has been criticized as being imprecise, and supported as offering a possible means of obtaining at least imprecise estimates when no other method can be used at all. Concurrent with Hamilton's final corrections was the critique by Fager (1969), who objected primarily to the necessary assumption of linear growth. He suggested, instead, employing losses between successive samples as being more accurate, but this procedure is the equivalent of the older removal-summation method and requires distinction of cohorts. As such, it would not fulfill the original intent of the method, i.e. to allow an estimate for a species or a group of species in which cohorts could not be distinguished.

Recently, Zwick (1975) also raised objections based on non-linear growth, attempting to show by re-working some of Hamilton's tables of hypothetical populations that non-linear growth has a very substantial effect on the result. However, Zwick apparently misinterpreted the tables of Hamilton's such that his conclusion of a serious discrepancy was itself in error (A. C. Benke and A. L. Hamilton, pers. comm.; Benke and Waide, 1977). With the correct interpretation, Hamilton's conclusion that irregular growth has little serious influence on the total estimate appears to remain valid.

Several empirical tests have now been completed, comparing estimates by the Hynes method with those by older methods, on single species populations, with favorable results (Hudson and Swanson, 1972; Neveu, 1973; Waters and Crawford, 1973; Cushman *et al.*, 1975), even where non-linear growth was included. Many other investigators have recently employed the method (Eckblad, 1973; Winterbourn, 1974; Castro, 1975; Erman and Erman, 1975; Horst and Marzolf, 1975; McClure and Stewart, 1976). Horst and Marzolf have presented a model in equation form for making production computations by the Hynes method. The only recent attempt to apply the Hynes method to a multi-species population apparently is that of Fisher and Likens (1973), who estimated the annual production of the aquatic insect fauna in Bear Brook, part of the New Hampshire Experimental Forest studies. All of these papers were concerned with benthic macroinvertebrates. Apparently, the Hynes method has not been attempted as yet with fish or zooplankton, although there is no fundamental reason why it cannot, provided the required raw data are obtained.

Zwick also questioned the proffered advantage of the method as applied to an entire fauna when it must be fractioned into subgroups by voltinism and growth pattern. However, while such fractioning may be difficult, it still may allow the estimate of a whole fauna not possible otherwise. Since the accuracy of the Hynes method has been indicated on single species, there seems little reason to believe it cannot be used for larger groups as well, provided the appropriate separations by trophic level, voltinism and maximum size are in fact made. Zwick's conclusion that the method is "wrong in principle" seems hardly justified.

In summary, to employ the Hynes method on an entire fauna, it is necessary to sort the organisms into several fractions: first, by trophic level into at least herbivore + detritivore *versus* predator groups; secondly, by voltinism at least into multivoltine, univoltine and hemivoltine; and thirdly, by maximum size. Obviously, some knowledge of invertebrate natural history and taxonomy is called for. The last of the three separations—by maximum size—will probably be the most difficult. It may be reasonably accurate to combine the species of a genus, or even of a family, and then decrease the "times loss" factor by 2 or 3 (from, say, 10 or 15) as a correction for the possibility that several species of varying maximum size were included or that the sexes were of different maximum sizes; while somewhat arbitrary, such a correction would at least be of the correct sign. Another possible error, with similar effect, may be introduced by basing the maximum size selection on an anomalously large individual (Waters and Crawford, 1973); it would appear more correct to take as the maximum size that of the largest size group containing substantial numbers.

F. PRODUCTION OF NON-COHORT POPULATIONS

With the exception of the Hynes method, the above methods estimate the production of discrete cohorts in their basic application. Many freshwater animals, however, do not occur in cohorts that are readily distinguishable, but rather with more-or-less continuous reproduction, and therefore those methods that estimate the production of a cohort are not applicable. Most zooplankters—rotifers and micro-crustaceans, primarily—fall into this group with continuous reproduction.

A few plankters, some calanoid copepods and others in regions of high latitude where low mean temperatures prevail, conform to a univoltine life history or at least have a small enough number of generations per year so that they may be distinguished and followed through the season as separate cohorts. In these cases, the methods outlined in previous sections for cohorts may be employed (Konstanti-nova, 1961; Yablonskaya, 1962; Comita, 1972; Rigler and Cooley, 1974; Anderson, 1975).

Since the main limitation with non-cohort populations is in the determination of growth rate, this parameter must be estimated with laboratory culture (e.g. Sorgeloos and Persoone, 1973); these laboratory-derived growth rates are then applied to standing stock data obtained from field samples. Since growth rates are dependent on water tempera-ture, the laboratory-derived data must be obtained at field-ambient temperatures or, more effectively, over a series of temperatures so that temperature-specific growth rates may be established for use at all field temperatures encountered (Geiling and Campbell, 1972). Alternatively, caged or bottled specimens may be cultured by immersion in the actual lake waters and thus at natural temperature and light conditions during a given production study (Korimek, 1966). While it is generally considered that developmental rate is mainly a function of temperature, food availability also may be effective in post-embryonic stages. Food undoubtedly also affects the mean weight of post-embryonic stages. This factor remains a problem for this group, not completely resolved.

The treatment of data then is basically an increment-summation method, in which the developmental time for each instar, group of instars, or size class is determined experimentally and, combined with measured or calculated weight increment for each stage, the mean daily weight increment for each stage is calculated. Daily pro-duction for each stage is calculated as the product of mean daily weight increment and standing stock in numbers, as determined from field samples, and summed for all stages. For annual production, the daily production estimates may be plotted over a year, and the annual total measured planimetrically.

The above procedure is only a basic outline; specific details of technique and equipment will depend upon life history characteristics of the animals involved and particular elements of the populations and the water body. For example, with copepods, the several stages of eggs, naupliar and copepodid instars may all be separately treated, or all the naupliar instars may be treated together, as may all copepodites. Cladocerans, without identifiable instars, may be separated by size groups. The method is usually assigned, in its most applicable form, to Winberg *et al.* (1965); criticized by Zawislak (1972), the method was defended by Ivanova (1973) and has been used widely.

The use of turnover rate, or "coefficient of renewal" (Elster, 1954), as derived from birth rates, was employed in some early studies of zooplankton production (Edmondson, 1960; Stross *et al.*, 1961; Wright, 1965). In this procedure, the finite birth rate, B, is computed as:

$$B = \frac{E}{D}$$

where E = standing stock of eggs in numbers (eggs/female × number of females) from field samples;

   D = development time of the embryos in days, determined experimentally in the laboratory.

Then the instantaneous birth rate, b, is computed as: $b = \ln{(B+1)}$. The reciprocal of b is then taken as the turnover time in days, and b as equivalent to turnover rate in percentage per day.

Daily production, then, is calculated as the product of b and mean standing stock. In this sense, it is implied that the daily turnover rate, b, is analogous to the daily P/B ratio, daily instantaneous growth rate, or Zaika's (1973) "specific production rate."

The computation of instantaneous birth rate, b, in the above fashion recently has been criticized on the basis that it requires the unrealistic assumption of zero mortality (Caswell, 1972), and Edmondson (1972) generally concurred. Paloheimo (1974) also criticized the method and provided an alternative means of computing b. Edmondson (1974) has included in his recent review an extensive analysis of limitations and assumptions in calculating birth rates. Furthermore, since turnover rates calculated from birth rates and development times are in numbers, it is inaccurate to equate these to turnover rates in biomass, unless steady state conditions obtain (Winberg, 1971), so that this turnover rate method would seem to have limited applicability.

In a different technique, Hall (1964) determined r, the instantaneous rate of population change, in laboratory cultures at different temperatures and food levels. The difference between calculated populations based on r, and observed populations in the field, was taken as the loss rate, which is equivalent to turnover rate, or the reciprocal of turnover time.

It should be emphasized that an accurate turnover time is not equivalent to the duration of life span; it will probably be about 0·2 times life span (based on an assumed cohort P/B ratio of about 5—see section on P/B ratios, p. 111). Juday (1940) and Lindeman (1941) essentially equated turnover time to life span (assumed cohort $P/B = 1$) and thus underestimated production.

A suggested alternative approach of obtaining daily turnover rates for zooplankters, which are not dependent on birth rates, but which are turnover rates in biomass, is as follows:

(1) Stage-specific instantaneous growth rates are computed as:

$$G = \ln \frac{\text{mean weight of stage}_{i+1}}{\text{mean weight of stage}_i}, \text{ where}$$

mean weights of stages are obtained from field samples and thus in accordance with actual food levels;

(2) Stage duration in days, D, is determined in laboratory culture at ambient temperature or over a series of temperatures;

(3) Daily turnover rates ($\approx$ daily instantaneous growth rates) for each stage are computed as:

$$\text{Daily turnover rate} = \frac{G}{D}$$

(4) Daily production is computed as the product of mean standing stock and daily turnover rate, for each stage, and all stages summed for population production.

Similarly, an approximate first estimate of production for a species population as a whole would include dividing the cohort instantaneous growth rate (assumed as 5, or calculated as $\ln \frac{\text{maximum weight}}{\text{initial weight}}$) by the cohort life span in days, determined in laboratory culture, to obtain a mean daily turnover rate applicable to the whole species population. Daily production is then computed as the product of this daily turnover rate and the mean standing stock of the whole species population.

While no one apparently has thus employed the instantaneous growth rate method for non-cohort planktonic crustaceans—although turnover rates, as used in this context, are analogous to instantaneous growth rates—it would seem that this would be an effective method. These procedures suggested above merely amount to obtaining daily instantaneous growth rates in laboratory culture and applying them to standing stock data in a manner similar to those described in the section on the instantaneous growth rate method.

For rotifers, the production rate may be taken as equivalent to finite birth rate, since the newly hatched individuals are already at or approaching their maximum weight (Edmondson, 1946). Edmondson (1960, 1968, 1971) presents the basic method as:

$$B = \frac{E}{D}$$

where B = finite birth rate in number eggs/female/day
E = egg ratio in eggs/female
D = embryonic development time in days

Duration of development time is determined by laboratory culture. The numbers produced per female per day are converted to numbers per unit volume by numbers of females per unit volume, from field samples. Numbers are converted to weight units from specific data on mean individual weight.

In addition to zooplankters, some benthic macroinvertebrates also occur in non-cohort populations. Generally, the same methods for non-cohort zooplankton may be applied to the zoobenthos, e.g. see Cooper (1965) who worked with the production of the amphipod, *Hyalella azteca*.

Raw field data for all planktonic groups are normally obtained per volume units, e.g. per cubic meter or per liter. In order to relate zooplankton production rates to other trophic levels or rates of energy supply, or to other water bodies, it is necessary to convert these data to area units, such as per square meter. If the volumetric units are for mean data in the water column, this is done by multiplying by the mean depth; if the data are from given strata with specific temperatures, they must be related to the same temperatures in the laboratory cultures; if the lake is small, or if there is a large ratio of littoral area to total area, the depth of specific strata must be weighted appropriately. Such factors as important seiches, differential currents at various depth, and vertically migrating zooplankters present problems, largely unresolved, that will cause difficulties in extrapolating estimates to the whole lake. Nevertheless, it is important in making the comparisons listed above to somehow express the production rates on an area basis.

Obtaining mean individual weight of zooplankters creates a much more difficult problem than for benthic macroinvertebrates. In the case of larger plankters, such as some cladocerans, individuals may be sorted out into size groups and actually weighed. With very small specimens, such as rotifers and most copepods, weight will have to be obtained indirectly, by calculating volume from approximate geometric shapes such as cylinders, ellipsoids, etc., using a specific gravity of about 1·05 (Hynes, 1961). Chissenko (1968) gives nomograms for determining water content. Some tables have been published for a variety of organisms, relating length measurements to weight (Edmondson and Winberg, 1971, p. 141, Dumont, 1975 and see references given by Edmondson, 1974, Appendix B).

Earlier papers in which non-cohort production rate methods for crustaceans were pioneered include Elster (1954), Stross et al. (1961), Hall (1964), Pechen and Shushkina (1964), and Wright (1965). General discussions of methods are presented by Edmondson and Winberg (1971), Winberg (1971), and Edmondson (1974).

G. ESTIMATING YIELD FROM PREDATOR CONSUMPTION

In earlier years, when direct production estimation methods for invertebrates were generally unavailable, minimal estimates of benthic production were computed as the rate of consumption by fish. These were closer to estimates of yield of the invertebrate level, rather than production, although the inference was that production would at least equal yield and undoubtedly be somewhat greater. Nevertheless, most of these early estimates of yield, back-calculated from fish production rates and food utilization efficiencies, indicated extremely high ratios of yield to mean biomass. Judged as a minimal estimate of the P/B ratio, they appeared unreasonably high (Allen, 1951; Hayne and Ball, 1956; Horton, 1961; Gerking, 1962); this has been termed "Allen's paradox" by Hynes (1970) (who consequently suggested that the fish were better samplers than the biologists!) after the very high ratios (up to 100) reported by Allen. Although some corrections were later made in Allen's results that reduced the ratio (Gerking, 1962), the ratios remained high.

There are various possible explanations for this paradox, almost all of which either underestimate the invertebrate standing stock or overestimate consumption by fish. However, the problem of comparing fish food consumption, as a measure of minimal benthic invertebrate production, with benthos standing stock is probably moot now, because modern method applications allow direct estimates of invertebrate production, and such "back-calculations" will not be necessary. Those few more recent cases wherein both fish production and invertebrate production have been estimated with direct methods have shown a more reasonable relationship.

We may continue to recall "Allen's paradox" as a reminder of the pitfalls that may beset researchers in production biology.

H. THE PHYSIOLOGICAL METHOD

The basis of the physiological method lies in knowledge of metabolic rates and the efficiency of assimilated energy used for growth, information obtained in respirometric studies in the laboratory on animals captured from the aquatic habitat of interest. The rates and efficiencies are specific to water temperature and to the individual's size. Combined with data on age- and size-specific standing stocks, field temperatures, and calorific content of the animals, estimates of the rate of energy used for growth in the entire population (= production) may be calculated.

Winberg (1971) presents detailed procedures in use of the physiological method; an outstanding example of the application of the method is that of Mann (1964, 1965) who studied energy flow in the fish population of the River Thames, England.

In view of the great detail of both the laboratory experiments and field sampling required, it would seem that the more empirical methods described above would in most cases be the more desirable, when production rate alone is the objective. The physiological method would be most applicable when the total energy budget of a population is desired, including the rates of energy flow in food ingestion, assimilation, respiration, etc., in addition to production.

IV. THE P/B RATIO

Much interest has been expressed in the ratio of production to standing stock, the P/B or turnover ratio.[1] It has long been observed that, for given groups of organisms and with a uniform computation, the P/B ratio is reasonably constant. Thus it has appeared to offer a more rapid, though perhaps less precise, means of arriving at estimates of production: multiplying a known P/B ratio by an appropriate measure of standing stock. In fact, some production data occurring in the literature seem to have been obtained in this way.

The P/B ratio may be computed in several ways, but it derives its basic significance from an expression as a *cohort* P/B: the total production of a cohort divided by the mean standing stock during the life span of the cohort, irrespective of the duration of the cohort's life span. Computed this way, the cohort P/B ratio appears reasonably

[1] The term *turnover ratio*, as used in some previous papers (e.g., Waters, 1969), seems inappropriate when applied to all types of animals, because it implies a certain recycling of materials or replacement of individuals and populations. Such is obviously not the case within the life span of a cohort. The term P/B, which is more explicit and also has an earlier usage, is employed throughout this review.

constant for almost all species, ranging from about 4 to 6 (Lawton, 1971; Jónasson, 1972; Waters and Crawford, 1973; Hunter, 1975; McClure and Stewart, 1976; Momot and Gowing, 1975, 1977).

It has been observed that the cohort P/B bears a close mathematical relationship to the cohort instantaneous growth rate (Waters, 1969; Mathews, 1970). It follows from the instantaneous growth rate formula for production:

$$P = G\bar{B}$$

that:

$$G = \frac{P}{\bar{B}}$$

Thus, Mathews (1970) suggested that at least a first approximation to a cohort's production would be to measure the maximum and minimum size of the animals in a cohort, calculate G as $\ln \dfrac{\text{maximum weight}}{\text{minimum weight}}$, and multiply G by a measure of $\bar{B}$, the mean standing stock over the cohort's life span.

Since cohort G is a function of the ratio, maximum size attained: size of newborn individuals, the cohort G, and thus the cohort P/B ratio, will fall within a range dictated by the range of the maximum/minimum ratio—e.g. 2·3 for a max/min ratio of 10, 4·6 for a max/min ratio of 100, and 6·9 for a max/min ratio of 1000. The max/min ratio of most freshwater animals appears to fall within the range of 10–1000, most usually 100–500.

The types of life history and growth pattern also have effects on the cohort P/B ratio, even for a given cohort G, although these variations appear to be relatively small. Tests with hypothetical Allen curves indicated that growth pattern (exponential, logarithmic, linear) and the relative size in numbers of the population at the end of the life span resulted in deviations of cohort P/B ratio from the cohort G (Waters, 1969). For example, species reaching the end of cohort life span with high numbers at approximately the same size (smolting salmonids ending their stream life, emerging amphibiotic insects) have a cohort P/B ratio smaller than their cohort G.

However, the most common expression of the P/B ratio, and probably the most useful ecologically, is that calculated on an annual basis. It is calculated as the annual production divided by the mean annual standing stock, irrespective of voltinism or life span, and may be calculated for a population of mixed ages or mixed species. The annual P/B ratio appears constant only for groups of animals with the same voltinism. In fact, the annual P/B depends primarily upon voltinism. Thus, for univoltine species, the annual P/B ratio will be nearly the same as the cohort P/B; for bivoltine species the annual P/B ratio will be roughly twice as large; for hemivoltine species it will be about one-half as large; etc.

The first clear expression of this dependence of the annual P/B ratio upon voltinism was that of K. H. Mann (1967) who predicted for benthos that the annual P/B would be about 10 for multivoltine species, 5 for univoltine, and 2 for species with a 2-year life span. These first approaches to a systematic organization of annual P/B ratios have proven remarkably accurate with the accumulation of empirical data.

It should be noted that the effect of voltinism upon annual P/B ratios is not exact. For example, for a univoltine aquatic insect, the annual P/B will be somewhat higher than the cohort P/B. Whereas the annual P will be the same as cohort P, annual $\bar{B}$ will be smaller than cohort $\bar{B}$ because the animals will be absent from the standing stock samples for a certain portion of the year (in egg stage, diapause, pupation, flying adult, etc.). The effect in this case is to make the annual P/B larger than the cohort P/B.

The modes of published annual P/B ratios in the literature appear to be as follows: multivoltine zooplankton (probably about 4–5 generations per year): 20; univoltine zoobenthos: 5; multivoltine zoobenthos (2–3 generations per year): 10; hemivoltine zoobenthos (2–3 years life span): 1·2; stream salmonid fishes (effective life span about 3 years): 1·2; warm-water fishes (life span to 10 years or more): 0·6. In the last three cases, there are overlapping cohorts, not successive, separated generations. (See data in succeeding section.)

It would appear that annual production estimates with sufficient precision for resource management decisions could probably be made employing the above modes of annual P/B and a reasonably good measure of mean standing stock. However, a few exceptional circumstances should be noted.

First, when annual P/B ratios are calculated for successive portions of a cohort's life span, it is observed that the ratio is higher during the earlier periods, for example during the first and second years of a fish year class (Hunt, 1966); similarly, annual P/B ratios during the juvenile life of migratory salmonids are higher (Chapman, 1965; Hopkins, 1971). This is because growth rates are high in the immature stages, as is true with most animals.

Secondly, a population in an expanding or colonization stage will show a higher annual P/B ratio, such as the stream trout population recovering from flood damage reported by Hanson and Waters (1974), because growth is high relative to mortality.

Third, a population that is overcrowded and stunted, such as is frequently observed with centrarchid populations in temperate lakes, will probably show a lower annual P/B ratio. A longer life span is indicated in this case, being the result of slower growth.

Fourth, annual P/B ratios for zooplankton in cold, northern or high altitude lakes have been found to be lower than the 20 suggested above (Winberg et al., 1973; Anderson, 1975; Rey and Capblancq, 1975). The reason is the slow growth and the occurrence of low number of generations per year compared with the same species at lower latitudes.

Fifth, it has been suggested by Johnson and Brinkhurst (1971) that annual P/B ratio is a function of water temperature and that P/B is approximated by $\dfrac{T^2}{10}$, where $T = $ mean °C at bottom of the lake. While their data (and many others') appear to confirm this relationship, a more direct reason is that water temperature affects voltinism. For example some species of Chironomidae, Simuliidae, Ephemeroptera, Crustacea and perhaps other taxa exhibit fewer generations per year at lower temperatures.

Recently, a number of researchers, mostly French, have begun to express the ratio as P/B_{max}, i.e. annual production divided by the maximum standing stock observed during the year (Laville, 1971, 1975; Giani and Laville, 1973; Décamps and Lafont, 1974; Potter and Learner, 1974; Lavandier, 1975; Otto, 1975; Rey and Capblancq, 1975). They report the P/B_{max} ratio to be fairly constant at about 1·5, for lake zooplankton and benthos in both lakes and streams, irrespective of voltinism. If truly constant, this ratio would prove invaluable in production biology studies, although conceivably it might be more difficult to determine B_{max} than $\bar{B}$ during a year. More empirical data for comparison are desirable.

In a small but data-packed book, Zaika (1973) presents many examples of *daily* P/B ratios, which he terms "specific production", and computes by dividing daily production by standing stock on the same day. Zaika suggests that specific production varies with species, can serve as an indicator of environmental conditions, and may be manipulated artificially for human benefit. It is, of course, conceptually similar, and numerically very close, to daily instantaneous growth rate. For the individual organism, specific production increases with temperature, decreases with age, and is generally smaller with longer-lived species, all of which should be predictable from present general knowledge of P/B ratios. Predation pressure maintains a high specific production in a prey population by cropping the large and older individuals and keeping a high proportion of young individuals in the prey population. When a population is increasing rapidly, the specific production is similar to r, the intrinsic rate of natural increase. Species-specific values are given in Zaika's book for rotifers, planktonic and benthic crustaceans, chironomids and molluscs.

A novel concept, similar to the P/B ratio, has been put forth as the P/E ratio—production : emergence—for amphibiotic insects, by Speir and Anderson (1974). They indicated a fairly constant P/E ratio for several simuliids in Oregon streams, most ranging from 4–5. The authors suggested the possibility of employing this statistic in estimating production rates.

Illies (1975) has also considered the use of emergence data toward the estimation of production rate of stream insects.

V. Annual Production in Secondary Producer Groups

Many estimates of annual production for the various secondary producer groups are now available, enough, in fact, so that we have some idea of general levels to be expected. The forms in which these data have been reported, however, are almost as numerous as the number of researchers; for purposes of summarization into the following tables and figures, it was necessary to convert to common units. In accordance with what seemed to be the most common usage of units, production rates of fishes are herein reported in kg/ha/yr (wet weight), of zoobenthos in kg/ha/yr (dry weight), and of zooplankton in g/m²/yr (dry weight). Data reported in other units were converted with the conversion coefficients in Table I, when coefficients were not given by the author.

In many excellent reports of zooplankton production, data were given in terms of volume as the spatial unit—e.g. g/m³/time (Zhdanova,

TABLE I

*Conversion coefficients employed in this review to express production rates in common units.**

Fish
1 g wet weight = 0·2 g dry weight
1 g wet weight = 1 kcal
Invertebrates (zoobenthos and zooplankton)
1 g dry weight = 6 g wet weight
1 g dry weight = 5 kcal
1 g dry weight = 0·9 g ash-free dry weight
1 g dry weight = 0·5 g Carbon

* These coefficients are neither absolute constants nor precise means; rather, they were used as what appeared to be approximate modes for convenience.

1969; Stepanova, 1971; Schindler, 1972; Smyly, 1973; Hart and Allanson, 1975; Lair, 1975; LaRow, 1975); unfortunately, without information on the productive depth of the water body, it was impossible to use these data for comparison with other reports giving production rates on an area basis. An occasional report gave zooplankton production using a daily or other time interval, without information on the length of the effective growing season (Coffman *et al.*, 1971; Duncan, 1975), in which case it could not be compared with annual rates.

When it is desired to compare production rates within an entire ecosystem, the calorie is the obvious unit to be used, and while many recent authors give data in calorific terms, the unit is not universally used. For most precise results, it is best to make direct calorific determinations of the organisms being studied. However, this is probably too much to expect from all institutions and laboratories, particularly since caloric content appears to vary with life stage, sex, age and environmental conditions. Nevertheless, caloric content seems to be reasonably constant with taxonomic groups, and a number of workers have provided data and tables from which at least approximate values can be obtained (Golley, 1961; Moshiri and Cummins, 1969; Platt *et al.*, 1969; Cummins and Wuycheck, 1971; Schindler *et al.*, 1971; Driver *et al.*, 1974; Caspers, 1975a, b). Methods for measuring caloric content of aquatic animals have been reviewed by Paine (1971) and Richman (1971).

In this review dealing with secondary production in inland waters, data from marine environments are of course not included, except where certain works are basically or historically significant to the development of methods. Much marine work has been done, particularly with zooplankton. Reviews of marine secondary production have been presented by Mullin (1969), Crisp, D. J. (1971, 1975), Holme and McIntyre (1971).

A. FISH

Data are accumulating rapidly on annual production of fish in a wide variety of environments (Tables II, III, Fig. 2). Outstanding among these data are the studies on the fishes of the River Thames, near Reading, England, in which extensive efforts were applied to obtain estimates for all elements of the fish fauna (Mann, K. H., 1964, 1965, 1971; Mathews, 1971; Berrie, 1972; Mann, K. H. *et al.*, 1972).

The River Thames studies included estimates among the highest so far reported, for both the entire fauna and a single species—perhaps partly because of the completeness of the work—approaching 2000

TABLE II

Single-species and total fauna estimates of annual production (kg/ha—wet weight) and annual P/B ratios for fishes (except stream salmon and trout). NA = data not available.

Species	Annual production	P/B	Locality	Remarks	Authority
Salmonidae (lentic waters)					
Salvelinus fontinalis	6·7	NA	New York lakes	Ave. 4 lakes, 2 yrs	Hatch and Webster (1961)
Salvelinus fontinalis	2·2	0·71	Matamek Lake, Quebec		Saunders and Power (1970)
Salvelinus fontinalis	0·21	0·62	Bill Lake, Quebec	Small populations, preyed by eels	O'Connor and Power (1973)
Salmo gairdneri	47·6	2·0	Whistler's Bend Imp., Oregon	Juveniles	Coche (1967)
Oncorhynchus nerka	66	NA	Cultus Lake, B.C.	Juveniles, max. of several yrs	Ricker and Foerster (1948)
Cottidae					
Cottus gobio	8·5	NA	River Tees tribs., England		Crisp, D. T. et al. (1974)
Cottus gobio	74·3	NA	River Tees, England		Crisp, D. T. et al. (1975)
Cottus gobio	172	NA	Bere Stream, England	Ave. 3 sites	Mann, R. H. K. (1971)
Cottus gobio	431	NA	River Tarrant, England		Mann, R. H. K. (1971)
Cottus gobio	144	NA	Devil's Brook, England		Mann, R. H. K. (1971)
Cottus cognatus	59·4	1·2	Valley Cr., Minn.		Petrosky and Waters (1975)

E

TABLE II (*continued*)

Species	Annual production	P/B	Locality	Remarks	Authority
Cottus carolinae	77·0	5·5	Steeles Run, Ky.	2nd Order stream	Small (1975)
Cottus carolinae	8·0	1·8	Steeles Run, Ky.	3rd Order stream	Small (1975)
Percidae					
Perca fluviatilis	0·04	0·41	Vistula R., Poland		Backiel (1971)
Perca fluviatilis	7·3	0·35	Reservoir, Czechoslovakia	Ave. 3 yrs	Holcik (1972)
Stizostedion lucioperca	0·91	0·72	Vistula R., Poland		Backiel (1971)
Stizostedion lucioperca	5·5	0·57	L. Balaton, Hungary		Biró (1975)
Stizostedion vitreum vitreum	2·1	0·43	West Blue L., Manitoba		Kelso and Ward (1972)
Stizostedion vitreum vitreum	2·8	NA	West Blue L., Manitoba		Ward and Robinson (1974)
Etheostoma spectabile	52	2·4	Steeles Run, Ky.	2nd Order streams	Small (1975)
Etheostoma flabellare	23	1·7	Steeles Run, Ky.	3rd Order streams	Small (1975)
Esocidae					
Esox lucius	14·2	NA	L. Windermere, England	For year 1951	Kipling and Frost (1970)
Esox lucius	1·5	0·7	Vistula R., Poland		Backiel (1971)
Esox lucius	0·75	0·32	Reservoir, Czechoslovakia		Holcik (1972)
Cyprinidae					
Aspius aspius	1·54	0·55	Vistula R., Poland		Backiel (1971)
Leuciscus cephalus	0·10	0·43	Vistula R., Poland		Backiel (1971)
Leuciscus cephalus	2·1	0·27	Reservoir, Czechoslovakia		Holcik (1972)
Tinca tinca	0·17	0·18	Reservoir, Czechoslovakia		Holcik (1972)
Rutilus rutilus	19·2	0·24	Reservoir, Czechoslovakia		Holcik (1972)
Rutilus rutilus	281	1·12	R. Thames, England	Max. 2 yrs	Mathews (1971)

Alburnus alburnus	915	1·92	R. Thames, England		Mathews (1971)
Leuciscus leuciscus	52	1·75	R. Thames, England		Mathews (1971)
Gobio gobio	289	1·94	R. Thames, England		Mathews (1971)
Others					
Lepomis macrochirus	91·2	NA	Wyland L., Ind.		Gerking (1962)
Philypnodon breviceps	39·8	NA	Small Spectacles L., N.Z.		Staples (1975)
Total fish fauna, Multi-species					
20 species	1306	NA	L. Kariba, Central Africa	Basis of entire lake	Balon (1974)
20 species	3468	NA	L. Kariba, Central Africa	Basis of fish-inhabited area only	Balon (1974)
Mostly *R. rutilis*	970	NA	Klicava Reservoir, Czechoslovakia	Maximum in reservoir development	Holcik and Pivnicka (1972)
Cyprinidae, Percidae, Catostomidae	89·4	NA	Clemons Fork, Ky.	Average of 1st, 2nd and 3rd order stream	Lotrich (1973)
Mostly Cyprinidae	1980	NA	R. Thames, England		Mann, K H. *et al.* (1972)
All trophic levels	244	0·85	Rybinsk res., U.S.S.R.		Sorokin (1972)
Planktivores	22	0·7	Naroch L., U.S.S.R.		Winberg *et al.* (1972)
Benthivores	13	0·5	Naroch L., U.S.S.R.		Winberg *et al.* (1972)
Piscivores	9	0·4	Naroch L., U.S.S.R.		Winberg *et al.* (1972)
Planktivores	24	0·7	Myastro L., U.S.S.R.		Winberg *et al.* (1972)
Benthivores	26	0·4	Myastro L., U.S.S.R.		Winberg *et al.* (1972)
Piscivores	14	0·3	Myastro L., U.S.S.R.		Winberg *et al.* (1972)
Planktivores	9	0·8	Batorin L., U.S.S.R.		Winberg *et al.* (1972)
Benthivores	60	0·5	Batorin L., U.S.S.R.		Winberg *et al.* (1972)
Piscivores	14	0·4	Batorin L., U.S.S.R.		Winberg *et al* (1972)

kg/ha/yr (wet weight) for the whole fish fauna. It should be noted that the production of youngest fry were included in these estimates, these stages contributing a very high proportion to annual production (Mathews, 1971); the youngest fry stages are often missed in many of the other studies. Also probably contributing to high production was the enriched condition of the Thames.

Other high estimates included those in the Klicava Reservoir, Czechoslovakia (Holcik and Pivnicka, 1972), a large reservoir in its developmental stages.

Most noteworthy, however, is the fact that these high estimates included mostly production of herbivorous-detritivorous fishes, such as some of the large cyprinids like the roach (*Rutilus rutilus*) and bleak (*Alburnus alburnus*). It is probably unlikely that carnivorous fishes will ever be found to reach similar high levels.

Available single-species estimates of annual production of carnivorous, and especially piscivorous, species have been considerably lower; outstanding among these is the estimate of 431 kg/ha/yr for the sculpin, *Cottus gobio*, in the hard-water River Tarrant, England (Mann, R. H. K., 1971). Estimates for salmonids in northern, soft-water lakes have been low at less than 10 kg/ha/yr (Hatch and Webster, 1961; Saunders and Power, 1970; O'Connor and Power, 1973). Piscivorous fishes such as percids and northern pike, as probably should be expected, also showed very low annual production (Kipling and Frost, 1970; Backiel, 1971; Holcik, 1972; Kelso and Ward, 1972; Ward and Robinson, 1974; Biro, 1975); among these, the northern pike (*Esox lucius*) of Lake Windermere, England, showed the highest at 14·2 kg/ha/yr (Kipling and Frost, 1970).

In an attempt to approximate expected fish production from an analysis of environmental factors, Huet (1964) suggested a method employing coefficients based on some knowledge of the fish population, fish food, temperature, alkalinity of the water, etc. Several workers concerned with fish production in streams have experimented with this method (Mills, 1967; Bishop, 1973), although these results are not included in the summary tables.

Le Cren (1962), Backiel (1967) and Toetz (1967) have emphasized that the sex products produced by a fish population constitute a high proportion of annual production, which is frequently overlooked in many studies. This factor is variable, however, as Staples (1975) reports a very low proportion for the New Zealand upland bully (*Philypnodon breviceps*).

In contrast to most invertebrates, annual P/B ratios for fishes are generally lower, the result of the longer life spans of several years (Tables II, III). Many of these range well below 1·0, reflecting life spans of up to 10 years or more. However, where estimates are available for populations of young stages only, the annual P/B ratio is higher, such as that for juvenile *Salmo gairdneri* in an Oregon impoundment at 2·0 (Coche, 1967). The highest values for the large cyprinids in the

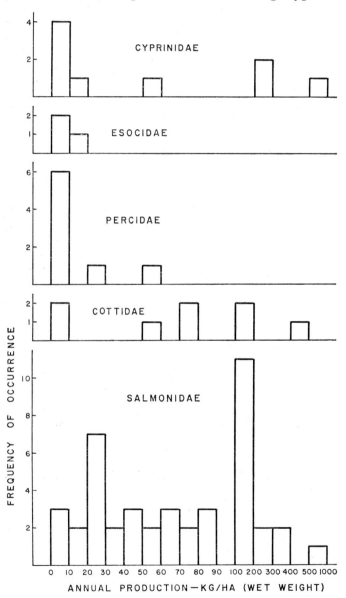

FIG. 2. Frequency occurrence of annual fish production estimates reported in the literature. Note non-linear abscissal scale.

TABLE III

Single-species and multi-species estimates of annual production (kg/ha—wet weight) and annual P/B ratios for salmon and trout in streams. NA = data not available.

Annual production	P/B	Stream and locality	Remarks	Authority
Salvelinus fontinalis				
300	NA	Big Spring Cr., Penn.		Cooper and Scherer (1967)
58	NA	Larry's Cr., Penn.		Cooper and Scherer (1967)
52	NA	Valley Cr., Minn.		Elwood and Waters (1969)
145	1·3	Valley Cr., Minn.	Flood damage, 2 yrs	Hanson and Waters (1974)
116	1·6	Lawrence Cr., Wisc.	Post-flood recovery, 3 yrs	Hunt (1974)
			Production, 11 yrs	Hunt (1969)
			P/B—6 yrs	
258	NA	Lawrence Cr., Wisc.	Max. 11 yrs	Hunt (1974)
152	NA	Lawrence Cr., Wisc.	Pre-habitat alteration, 4 yrs	Hunt (1974)
205	NA	Lawrence Cr., Wisc.	Post-habitat alteration, 7 yrs	Hunt (1974)
38	0·7	Gallienne Cr., Quebec	Ave. 2 yrs	O'Connor and Power (1976)
63	1·6	Kaikhosru Cr., Quebec	Ave. 2 yrs	O'Connor and Power (1976)
25	1·05	Sherry Cr., Quebec	Ave. 3 yrs	O'Connor and Power (1976)
16	1·2	Tehinicaman R., Quebec	Ave. 2 yrs	O'Connor and Power (1976)
Salmo trutta				
544	2·0	Horokiwi Stream, N.Z.	Ave. 6 sites	Allen (1951)
380	NA	Horokiwi Stream, N.Z.	Possible correction	Le Cren (1969)
106	1·0	Black Earth Cr., Wisc.	Polluted stream	Brynildson and Mason (1975)
328	1·2	Black Earth Cr., Wisc.	Pollution abated	Brynildson and Mason (1975)
66	NA	R. Tees tribs., England	4 streams, 3 yrs	Crisp, D. T. et al. (1974)
22	1·1	3 streams, England	Normal growth	Crisp, D. T. et al. (1975)

Species					Reference
	23	0·5	Great Dodgen Pot, England	Slow growth	Crisp, D. T. et al. (1975)
	101	NA	Shelligan Burn, Scotland	3 yrs	Egglishaw (1970)
	85	4·9	Hinau Main Water, N.Z.	Nursery stream	Hopkins (1971)
	89	5·0	Hinaki Main Water, N.Z.	Nursery stream	Hopkins (1971)
	132	1·1*	Walla Brook, England	2 yrs	Horton (1961)
	38	0·5	River Yarty, England	Fry produced in tribs.	Horton et al. (1968)
	78	NA	10 streams, England		Le Cren (1969)
	129	NA	Bere Stream, England		Mann, R. H. K. (1971)
	120	NA	R. Tarrant, England		Mann, R. H. K. (1971)
	48	NA	Devil's Br., England		Mann, R. H. K. (1971)
	121	NA	Docken's Water, England		Mann, R. H. K. (1971)
	26	NA	Streams in northern Norway	Ave. of maxima	Power (1973)
Salmo gairdneri	132	2·4	Bothwell's Cr., Ontario	Juveniles only	Alexander and MacCrimmon (1974)
	104	NA	Big Springs Cr., Idaho		Goodnight and Bjornn (1971)
	24	NA	Lemhi R., Idaho		Goodnight and Bjornn (1971)
	28	NA	Valley Cr., Minn.	Max. 3 yrs	Hanson and Waters (1974)
Salmo clarki	41	1·0	3 streams, Oregon		Lowry (1966)
Salvelinus alpinus	25	NA	Kromagelva R. system, northern Norway	Max. several streams	Power (1973)
Oncorhynchus kisutch	86	2·4	3 streams Oregon	4 yrs, juveniles only	Chapman (1965)

TABLE III (*continued*)

Annual production	P/B	Stream and locality	Remarks	Authority
Salmo salar				
72	NA	Bere Stream, England	Juveniles	Mann, R. H. K. (1971)
11	NA	streams in northern Norway	Ave. maxima	Power (1973)

Annual production	Salmonid species	Locality	Remarks	Authority
Multi-species salmonid populations				
127	Coho salmon, cutthroat trout	3 streams, Oregon		Chapman (1965) Lowry (1966)
195	Atlantic salmon, brown trout	Shelligan Burn, Scotland	Ave. 3 yrs	Egglishaw (1970)
110	Rainbow and brook trout, chinook salmon	Big Springs Cr., Idaho		Goodnight and Bjornn (1971)
54	Rainbow trout, chinook salmon	Lemhi R., Idaho		Goodnight and Bjornn (1971)
158	Brook and rainbow trout	Valley Cr., Minn.	Ave. 3 yrs	Hanson and Waters (1974)
60	Brown trout, sea trout, Atlantic salmon	R. Yarty, England		Horton *et al.*, (1968)
201	Brown trout, Atlantic salmon	Bere Stream, England		Mann, R. H. K. (1971)

* Part of Table 7 of Horton (1961), leading to the calculation of 11·4 for P/B is apparently in error by a factor of 10.

River Thames apparently reflect the great contribution by young stages as well as the high ratios of adult-to-fry weight (Mathews, 1971). Where available, P/B ratios for cottids appear to be higher, perhaps reflecting shorter life spans (Petrosky and Waters, 1975; Small, 1975).

The recent book on tropical Lake Kariba, adjoining Zambia and Rhodesia, an impoundment of the Zambezi River closed in 1958, includes production data for 20 of the more common species of fishes present in the reservoir (Balon, 1974). The total annual production estimated for these 20 species was 3468 kg/ha/yr for that part of the lake inhabited by fish, or 1306 kg/ha/yr on the basis of the whole lake. As such, the estimate for the fish-inhabited part of the lake represents by far the highest fish production rate yet published. Although a large, comparatively new reservoir, particularly in the tropics, would be expected to have a high production rate, these figures seem somewhat overestimated, particularly since mean standing stocks were not comparably high (i.e. annual P/B ratios were high at about 4, for these long-lived species) and since the species with the greatest calculated production were mainly carnivores. The production calculated for age $0+$ stages for the most productive species seemed excessively high (up to 98% of total) and may have been overestimated. Nevertheless, total fish production in this tropical reservoir is undoubtedly very high, and may still represent the world's freshwater type most productive of fish.

1. Stream trout and salmon

Of some particular importance is the considerable body of data now available on annual production of salmonids in small streams (Table III). The most extensive of these is the valuable series of 11 years' continuous data on brook trout (*Salvelinus fontinalis*) from Lawrence Creek, Wisconsin, reported by Hunt (1966, 1969, 1971, 1974). Studies on trout and salmon in streams undoubtedly have been enhanced by interest in recreational fisheries management in these streams, as well as by the development of highly efficient fish collecting methods using electrofishing apparatus.

Allen (1951) made the first estimates of annual production of stream trout, which remain the highest yet reported today; his estimates have been questioned as having systematic errors causing an overestimate, but Allen did not have the advantages of electrofishing apparatus. Even discounting his estimates, they will probably remain among the highest in this category.

Le Cren (1969) suggested, upon his review of data from 10 English streams, that a maximum of about 120 kg/ha/yr might apply in stream salmonid populations; additional data collected since his summary would suggest something higher than that as a maximum—perhaps about 300 kg/ha/yr. Data to date may also indicate that an annual production of around 100 kg/ha/yr, which might commonly be found in high-alkalinity streams (the chalk or limestone streams of anglers' parlance), would indicate a highly productive trout population capable of sustaining a reasonably intensive sport fishery. Most soft-water, northern streams appear to exhibit a much lower production—about 15–50 kg/ha/yr, e.g. the streams of the Canadian Shield reported by O'Connor and Power (1976), the northern Norwegian streams of Power (1973), soft-water Larry's Creek in Pennsylvania (Cooper and Scherer, 1967), and the soft-water streams in England (Crisp, D. T. *et al.*, 1975); the soft-water Walla Brook with an annual production of 132 kg/ha (Horton, 1961), appears as a striking exception.

Most P/B ratios, calculated on an annual basis for normal, stream-inhabiting populations are fairly regular at something slightly above 1·0 (Table III). (P/B ratios calculated from Allen's (1951) data are anomalous, perhaps related to possible overestimates already mentioned.) Where the salmonid populations consist almost entirely of juveniles, with their higher growth rates, the P/B ratios are higher, up to about 5: the brown trout nursery streams in New Zealand (Hopkins, 1971), juvenile migrating rainbows in Bothwell's Creek, Ontario (Alexander and MacCrimmon, 1974), and the juvenile coho salmon in Oregon streams (Chapman, 1965). On the other hand, low values appear where the population is heavy towards the older individuals, such as the brown trout in the River Yarty where spawning and fry production take place in the tributaries and the river population consists of older fish, with a P/B of 0·5 (Horton *et al.*, 1968); also, the low value of 0·5 for the brown trout in Great Dodgen Pot, England, was apparently due to extremely slow growth and an age distribution heavy with older, slow-growing fish (Crisp, D. T. *et al.*, 1975).

B. ZOOBENTHOS

Single-species estimates of annual production for benthic invertebrates vary greatly, as may be expected (Table IV, Fig. 3). The highest rates range up to several hundred kg/ha/yr (dry weight), with the highest being the 405 kg/ha/yr for the chironomid, *Glyptotendipes*, in Loch Leven, Scotland, reported by Maitland and Hudspith (1974). Kimerle and Anderson (1971) reported 1616 kg/ha/yr for *Glyptotendipes barbipes* in a narrow stratum near the shore of an Oregon sewage lagoon, where the larvae were living in great concentration; although based on the lagoon as a whole, the annual production was much less—74 kg/ha/yr. Other high estimates of over 100 kg/ha/yr were reported for

TABLE IV

Single-species estimates of annual production (kg/ha/yr—dry weight) and annual P/B ratios for zoobenthos.
NA = data not available.

	P	Voltinism	P/B	Locality	Remarks	Authority
Insecta						
Diptera						
Chironomidae						
Tanytarsus jucundus	12·5	Uni-	NA	Sugarloaf L., Mich.		Anderson and Hooper (1956)
Chironomus anthracinus	128·9	2 yrs	3·8	L. Esrom, Denmark		Jónasson (1972)
Procladius pectinatus	5·2	Uni-	1·9	L. Esrom, Denmark		Jónasson (1972)
Chironomus anthracinus	224·4	2 yrs	NA	L. Esrom, Denmark	Ave. 3 yrs	Jónasson (1975)
Microtendipes chloris	0·24	Multi-	12·7	Mikolajskie L., Poland		Kajak and Rybak (1966)
Glyptotendipes barbipes	74	Multi-	8·5	Waste lagoon, Oregon	Entire lagoon	Kimerle and Anderson (1971)
Glyptotendipes barbipes	1616	Multi-	8·5	Waste lagoon, Oregon	Nearshore stratum	Kimerle and Anderson (1971)
Psectocladius sordidellus	7·6	Uni-	NA	Lake Port-Bielh, France		Laville (1971)
Zavrelimyia melanura	2·7	Uni-	NA	Lake Port-Bielh, France		Laville (1971)
Chironomus commutatus	6·3	Bi-	NA	Lake Port-Bielh, France		Laville (1975)
Limnochironomus	5·9	Multi-	3·6	Loch Leven, Scotland	Max. 2 yrs	Maitland and Hudspith (1974)

TABLE IV (continued)

	P	Voltinism	P/B	Locality	Remarks	Authority
Glyptotendipes	405	Multi-	5·0	Loch Leven, Scotland	Max. 2 yrs	Maitland and Hudspith (1974)
Stictochironomus	102	Multi-	9·8	Loch Leven, Scotland	Max. 2 yrs	Maitland and Hudspith (1974)
Stictochironomus	118	Multi-	NA	Loch Leven, Scotland		Maitland et al. (1972)
Procladius choreus	34·1	Bi-	6·5	Eglwys Nunydd res., S. Wales, U.K.		Potter and Learner (1974)
Procladius rufovittatus	20·1	Uni-	3·2	Eglwys Nunydd res., S. Wales, U.K.		Potter and Learner (1974)
Glyptotendipes paripes	30·8	Bi-	5·9	Eglwys Nunydd res., S. Wales, U.K.		Potter and Learner (1974)
Limnochironomus pulsus	11·0	Bi-	6·3	Eglwys Nunydd res., S. Wales, U.K.		Potter and Learner (1974)
Chironomus plumosus	35·1	Bi-	6·4	Eglwys Nunydd res., S. Wales, U.K.		Potter and Learner (1974)
Microtendipes sp.	16·2	Bi-	5·7	Eglwys Nunydd res., S. Wales, U.K.		Potter and Learner (1974)
Parachironomus tener	1·1	Bi-	5·8	Eglwys Nunydd res., S. Wales, U.K.		Potter and Learner (1974)
Cladotanytarsus mancus	1·0	Bi-	5·2	Eglwys Nunydd res., S. Wales, U.K.		Potter and Learner (1974)
Tanytarsus lugens	24·0	Tri-	7·6	Eglwys Nunydd res., S. Wales, U.K.		Potter and Learner (1974)
Tanytarsus holochlorus	15·5	Bi-	6·6	Eglwys Nunydd res., S. Wales, U.K.		Potter and Learner (1974)
Tanytarsus inopertus	9·1	Bi-	5·2	Eglwys Nunydd res., S. Wales, U.K.		Potter and Learner (1974)

Five spp.	6·3	2–3 yrs	1·5	Char L., Cornwallis Is., Can.	Total, long-lived spp. in polar region	Welch (1976)
Simuliidae						
Simulium equinum	35·6	Tri-	NA	Bere Stream, England		Ladle *et al.* (1972)
Simulium ornatum	25·3	Quadri-	NA	Bere Stream, England		Ladle *et al.* (1972)
Simulium brevicaule	0·60	Multi-	6·2	Lissuraga R., France	Ave. 2 sites	Neveu (1973)
Simulium monticola and dorieri	1·12	Multi-	11·1	Lissuraga R., France	Ave. 4 sites	Neveu (1973)
Simulium ornatum	1·29	Multi-	12·9	Lissuraga R., France	Ave. 2 sites	Neveu (1973)
Prosimulium caudatum	4·7	Uni-	NA	Three streams, Oregon	Ave. 2 yrs	Speir and Anderson (1974)
Prosimulium dicum	5·3	Uni-	NA	Three streams, Oregon	Ave. 2 yrs	Speir and Anderson (1974)
Simulium arcticum	6·4	Uni-	NA	Three streams, Oregon	Ave. 2 yrs	Speir and Anderson (1974)
Simulium canadense	5·7	Uni-	NA	Three streams, Oregon	Ave. 2 yrs	Speir and Anderson (1974)
Culicidae						
Chaoborus punctipennis	0·12	Multi-	NA	Severson L., Minn.		Comita (1972)
Chaoborus flavicans	29·2	Uni-	1·7	L. Esrom, Denmark		Jónasson (1972)
Stratiomyiidae						
Hedriodiscus truquii	13·6	Bi-	NA	Ohanapecosh Springs, 2 springs (thermal), Wash.		Stockner (1971)
Ephemeroptera						
Hexagenia limbata	1·2	Uni-	4·4	Tuttle Cr. res., Kansas	Ave. 4 yrs	Horst and Marzolf (1975)

TABLE IV (continued)

	P	Voltinism	P/B	Locality	Remarks	Authority
Hexagenia limbata and bilineata	15·0	1·7 yrs	2·8	Lewis and Clark L., S.D. and Nebraska		Hudson and Swanson (1972)
Choroterpes mexicanus	2·5	Tri-	15·4	Brazos River, Texas		McClure and Stewart (1976)
Baetis bicaudatus	13·7	Bi-	NA	Temple Fk., Logan R., Utah	Ave. 2 yrs	Pearson and Kramer (1972)
Ephemera strigata	7·1	Uni-	2·3	Yoshino R., Japan		Tsuda (1972)
Baetis vagans	21	Tri-	9·7	Valley Creek, Minn.		Waters (1966)
Ephemerella subvaria	44·5	Uni-	5·8	Luxemburg Cr., Minn.		Waters and Crawford (1973)
Deleatidium sp.	42·3	Bi-	3·5	Selwyn R., N.Z.		Winterbourn (1974)
Baetis rhodani	8·9	Bi-	8·4	Trout streams, Czechoslovakia	Ave. 3 sites	Zelinka (1973)
Rhithrogena semicolorata	20·8	Uni-	8·4	Trout streams, Czechoslovakia	Ave. 3 sites	Zelinka (1973)
Ecdyonurus sp.	14·1	Uni-	8·4	Trout streams, Czechoslovakia	Ave. 3 sites	Zelinka (1973)
Trichoptera						
Agapetus fuscipes	6·2	Uni-	NA	Brietenbach, W. Germany	Tissue only	Castro (1975)
Agapetus fuscipes	8·0	Uni-	NA	Breitenbach, W. Germany	Including exuviae and secretions	Castro (1975)
Diplectrona modesta	0·98	Uni-	NA	Walker Branch, Tenn.	Total, 2 univoltine generations	Cushman et al (1975)
Micrasema difficile	2·6	Uni-	NA	Pyrenees streams, France	Ave. 2 yrs	Décamps and Lafont (1974)
Micrasema morosum	0·38	1½ yrs	NA	Pyrenees streams, France	Ave. 2 yrs	Décamps and Lafont (1974)

Cyrnus trimaculatus	0·38	Uni-	4·6	R. Thames, England		Mann, K. H. (1971)
Potamophylax cingulatus	7·7	Uni-	4·4	Stampen Stream, Sweden	Tissue only	Otto (1975)
Potamophylax cingulatus	9·1	Uni-	NA	Stampen Stream, Sweden	Including exuviae and secretions	Otto (1975)
Oligophlebodes sigma	28·8	Uni-	NA	Temple Fk., Logan R., Utah	Ave. 2 yrs	Pearson and Kramer (1972)
Athripsoides ancylus	0·099	Uni-	5·8	Brashears Creek, Kentucky		Resh (1975)
Plecoptera						
Alloperla mediana	0·17	2 yrs	2·8	Walker Branch, Tenn.		Cushman *et al.* (1975)
Capnioneura brachyptera	0·68	2 yrs	NA	Pyrenees stream, France	Ave. 2 yrs	Lavandier (1975)
Archynopteryx curvata	NA	Uni-	3·8	Sagehen Creek, Calif.	P in arbitrary units	Sheldon (1972)
Stenoperla prasina	20·8	Uni-	2·5	Selwyn River, N.Z.		Winterbourn (1974)
Odonata						
Pyrrhosoma nymphula	7·1	2 yrs	3·2	Ponds in England		Lawton (1971)
Megaloptera						
Sialis lutaria	5·1	3 yrs	NA	Lake Port Bielh, France		Giani and Laville (1973)
Sialis mitsuhashii	2·5	3 yrs	3·9	Lake Tatsu-Nama, Japan		Yamamoto (1972)
Hemiptera						
Corixa germari	21	NA	NA	Barbrook Reservoir, England		Crisp, D. T. (1962)

TABLE IV *(continued)*

	P	Voltinism	P/B	Locality	Remarks	Authority
Crustacea						
Isopoda						
Asellus aquaticus	27·8	2 yrs	2·0	L. Pajep Måskejaure, Sweden		Andersson (1969)
Asellus aquaticus	52·7	Uni-	2·0	L. Erken, Sweden		Andersson (1969)
Asellus aquaticus	1·38	Multi-	3·2	R. Thames, England		Mann, K. H. (1971)
Asellus aquaticus	9·4	Bi-	4·7	Eglwys Nunydd res., S. Wales, U.K.		Potter and Learner (1974)
Amphipoda						
Gammarus tigrinus	130	Uni-	6·2	Tjeukemeer Reservoir, The Netherlands		Beattie *et al.* (1972)
Hyalella azteca	19·3	Bi-	4·8	Sugarloaf L., Mich.		Cooper (1965)
Pontoporeia affinis	5·2	Uni-	4·1	L. Krasnoye, U.S.S.R.		Kuz'menko (1969)
Gammarus pulex	0·14	Multi-	6·7	R. Thames, England		Mann, K. H. (1971)
Crangonyx pseudogracilis	0·22	Multi-	5·8	R. Thames, England		Mann, K. H. (1971)
Crangonyx richmondensis occidentalis	3·5	1·5 yrs	2·0	Marion L., B.C.		Mathias (1971)
Hyalella azteca	11·5	Uni-	4·2	Marion L., B.C.		Mathias (1971)
Gammarus pseudolimnaeus	366	NA	NA	Cone Spring, Iowa		Tilley (1968)
Decapoda						
Orconectes virilis	34·5	3·5 yrs	2·3	West Lost L., Mich.	1962–1963	Momot (1967)
Orconectes virilis	28·3	3·5 yrs	NA	South Twin L., Mich.	1966 only	Momot and Gowing (1975)

Species				Location		Reference
Orconectes virilis	8·2	3·5 yrs	1·1	North Twin L., Mich.	Ave. 9 yrs	Momot and Gowing (1977)
Orconectes virilis	22·3	3·5 yrs	0·94	West Lost L., Mich.	1966–1967	Momot and Gowing (1977)
Mollusca (production units shell-free)						
Mya arenaria	116	2+ yrs	2·5	Peteswick Inlet, Nova Scotia	Intertidal estuary	Burke and Mann, K. H. (1974)
Macoma balthica	19·3	2+ yrs	1·5	Peteswick Inlet, Nova Scotia	Intertidal estuary	Burke and Mann, K. H. (1974)
Littorina saxatilis	32·5	2+ yrs	4·1	Peteswick Inlet, Nova Scotia	Intertidal estuary	Burke and Mann, K. H. (1974)
Lymnaea palustris	21·8	NA	NA	Bool's Backwater, Fall Cr., N.Y.		Eckblad (1973)
Physa integra	15·9	NA	NA	Bool's Backwater, Fall Cr., N.Y.		Eckblad (1973)
Gyraulus parvus	0·33	Multi-	NA	Bool's Backwater, Fall Cr., N.Y.		Eckblad (1973)
Physa gyrina	11	NA	NA	Madison River, Mont.		Gillespie (1969)
Gyraulus deflectus	33	NA	NA	Madison River, Mont.		Gillespie (1969)
Valvata humeralis	9	NA	NA	Madison River, Mont.		Gillespie (1969)
Pisidium compressum	30	NA	NA	Madison River, Mont.		Gillespie (1969)
Lymnaea palustris	121	NA	NA	Ponds, streams in N.Y.		Hunter (1975)
Pisidium casertanum	4·38	2 yrs	0·2	Lake Esrom, Denmark		Jónasson (1972)
Corbicula africana	106·3	1–2 yrs	2·9	Lake Chad, Africa	Max. several sites	Leveque (1973)
Bellamya unicolor	156·3	Uni-	5·5	Lake Chad, Africa	Max. several sites	Leveque (1973)
Melania tuberculata	154·5	1–2 yrs	4·4	Lake Chad, Africa	Max. several sites	Leveque (1973)

TABLE IV (*continued*)

	P	Voltinism	P/B	Locality	Remarks	Authority
Cleopatra bulimoides	119·0	1+ yr	3·5	Lake Chad, Africa	Max. several sites	Leveque (1973)
Planorbis contortus	0·017	Uni-	0·3	R. Thames, England		Mann, K. H. (1971)
Planorbis vortex	0·11	Uni-	6·5	R. Thames, England		Mann, K. H. (1971)
Ancylus lacustris	0·09	Uni-	3·3	R. Thames, England		Mann, K. H. (1971)
Viviparus viviparus	4·46	2–3 yrs	1·6	R. Thames, England		Mann, K. H. (1971)
Bithynia tentaculata	1·1	2–3 yrs	1·6	R. Thames, England		Mann, K. H. (1971)
Sphaerium corneum	2·85	2–3 yrs	3·5	R. Thames, England		Mann, K. H. (1971)
Limnaea pereger (*Lymnaea peregra?*)	5·02	Multi-	11·5	R. Thames, England		Mann, K. H. (1971)
Unio pictorum	8·8	10–12 yrs	0·14	R. Thames, England		Negus (1966)
Unio tumidus	3·3	10–12 yrs	0·13	R. Thames, England		Negus (1966)
Anodonta anatina	22·0	10–12 yrs	0·20	R. Thames, England		Negus (1966)
Unio pictorum	3·7	2+ yrs	0·15	Danube River pools, Romania		Tudorancea (1972)
Unio tumidus	3·4	2+ yrs	0·25	Danube River pools, Romania		Tudorancea (1972)
Anodonta piscinalis	4·9	2+ yrs	0·33	Danube River pools, Romania		Tudorancea (1972)
Annelida						
Oligochaeta						
Tubificidae						
Ilyodrilus hammoniensis	11·7	NA	0·7	Lake Esrom, Denmark		Jónasson (1972)
Potamothrix hammoniensis	10·4	4 yr	0·86	Lake Esrom, Denmark		Jónasson (1975)
Limnodrilus spp.	53·6	Multi-	12·5	Eglwys Nunydd res., S. Wales, U.K.		Potter and Learner (1974)

Oligochaetes	12–97	NA	5·5–9·3	Peat land, Calif.		Erman and Erman (1975)
Polychaeta						
Neanthes virens	90·4	2 yrs	1·6	R. Thames estuary, England		Kay and Brafield (1973)
Hirudinea						
Erpobdella octoculata	36·0	2 yrs	1·5	Wilfin Beck, England	Ave. 3 yrs	Elliott (1973)
Helobdella stagnalis	11·9	Bi-	3·0	Eglwys Nunydd res., S. Wales, U.K.	Max. 2 yrs	Learner and Potter (1974)
Helobdella stagnalis	8·7	Bi-	3·5	Eglwys Nunydd res., S. Wales, U.K.		Potter and Learner (1974)
Helobdella stagnalis	2·25	Multi-	4·8	R. Thames, England		Mann, K. H. (1971)
Erpobdella octoculata	3·4	2–3 yrs	3·5	R. Thames, England		Mann, K. H. (1971)
Glossiphonia heteroclita	0·17	Uni-	5·3	R. Thames, England		Mann, K. H. (1971)
Glossiphonia complanata	0·58	2–3 yrs	2·1	R. Thames, England		Mann, K. H. (1971)

the midge *Chironomus anthracinus* in the profundal of Lake Esrom, Denmark (Jónasson, 1972, 1975), *Stictochironomus* in Loch Leven (Maitland *et al.*, 1972; Maitland and Hudspith, 1974), the amphipods *Gammarus tigrinus* in a Dutch reservoir (Beattie *et al.*, 1972) and *G.*

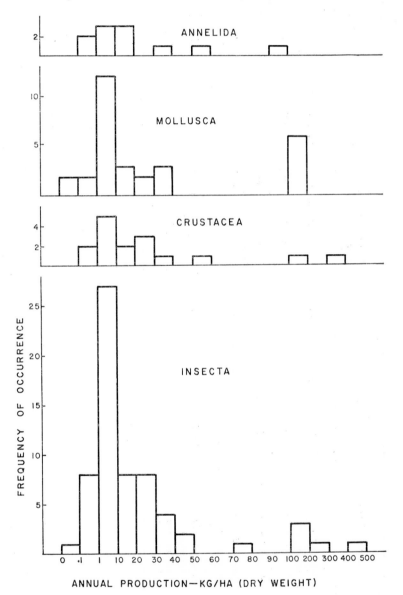

Fig. 3. Frequency occurrence of annual zoobenthos production estimates reported in the literature. Note non-linear abscissal scale.

pseudolimnaeus in a temperate spring brook (Tilley, 1968), the mollusc *Mya arenaria* in a Nova Scotian estuary (Burke and Mann, K. H., 1974), and for several species in Lake Chad, Africa (Leveque, 1973).

A few studies have attempted to estimate the annual production of the total zoobenthos fauna (Table V). Maxima appear to be well in the range of 500 kg/ha/yr and higher. The highest figure reported in Table V, 2000 kg/ha/yr, in the Speed River, Ontario, is the result of an approximation first calculated by Hynes and Coleman (1968) and roughly corrected by the present writer according to Hamilton's (1969) modification; while it appears higher than all other estimates reported, it seems reasonable in view of the inclusion of large quantities of benthic invertebrates recovered from the hyporheic region of the stream bottom (H. B. N. Hynes, pers. comm.).

Increasing concern for the inclusion in production estimates of molting losses and other materials not usually included has been expressed in the attempts of some workers. For example, Castro (1975) and Otto (1975) have estimated exuviae and case-building secretions of Trichoptera larvae, which constituted substantial proportions of the total.

Cook and Johnson (1974) have considered the possible general levels of benthos production in the Laurentian Great Lakes and suggested probable levels ranging from 0·1 $g/m^2/yr$ (wet?) in offshore Lake Superior to about 5 $g/m^2/yr$ in Lakes Michigan and Erie, with higher values in bays and nearshore areas.

Annual P/B ratios varied greatly among data reported, although most values followed the general rule of larger annual P/B ratios with increasing number of generations per year (Table IV, Fig. 4). Multivoltine dipterans frequently showed P/B ratios around 10—*Microtendipes chloris* (Kajak and Rybak, 1966), *Glyptotendipes barbipes* (Kimerle and Anderson, 1971), *Stictochironomus* (Maitland and Hudspith, 1974), *Tanytarsus lugens* (Potter and Learner, 1974), *Simulium* spp. (Neveu, 1973), as did also some multivoltine ephemeropterans—*Choroterpes mexicanus* (McClure and Stewart, 1976), and *Baetis vagans* (Waters, 1966). Most univoltine and bivoltine species, reported by many authors, most commonly showed P/B ratios in the neighborhood of 4–7, as did most of the crustacean isopods and amphipods. On the other hand, insects with life spans of over one year—chironomids in polar regions (Welch, 1976), *Hexagenia* mayflies (Hudson and Swanson, 1972), plecopterans (Cushman *et al.*, 1975), odonates (Lawton, 1971), decapod crustaceans (Momot and Gowing, 1975, 1977) and some annelids (Mann, K. H., 1971; Elliott, 1973; Kay and Brafield, 1973; Jónasson, 1975)—all showed lower P/B ratios from about 1–3. Most notable among long-lived species are some of the molluscs, with life spans up

TABLE V

Estimates of annual production (kg/ha/yr—dry weight) for zoobenthos, entire fauna.

	P	Locality	Remarks	Authority
Herb.-Detrit.	3·7	Lake Krivoe, U.S.S.R.		Alimov et al. (1972)
Herb.-Detrit.	9·0	Lake Krugloe, U.S.S.R.		Alimov et al. (1972)
Non-predator	38·8	Red Lake, U.S.S.R.		Andronikova et al. (1972)
Predator	2·0	Red Lake, U.S.S.R.		Andronikova et al. (1972)
Benthos and midges	84	Wyland Lake, Indiana		Gerking (1962)
All benthos	152	Marion Lake, B.C.		Hall and Hyatt (1974)
All benthos	2000	Speed River, Ontario	Approx., with Hamilton correction	Hynes and Coleman (1968)
All benthos	507	Bay of Quinte, Lake Ontario, Canada	Max. 4 sites, Ave. 2 yrs	Johnson and Brinkhurst (1971)
Non-predator	400	Polish lakes	Max. 5 lakes	Kajak et al. (1972)
Non-predator	204	Polish lakes	Ave. 5 lakes	Kajak et al. (1972)
All benthos	367	Lake Taltowisko, Poland		Kajak and Rybak (1966)
All benthos	283	Lake Mikolajskie, Poland		Kajak and Rybak (1966)
Dominant spp.	21	Lake Port-Bielh, France		Laville (1975)
Herb.-Detrit.	214·4	R. Thames, England		Mann, K. H. et al. (1972)
Predator	34·8	R. Thames, England		Mann, K. H. et al. (1972)
Herb.-Detrit.	556	Middle Oconee R., Georgia		Nelson and Scott (1962)
Herbivore	44·9	Volchja R. res., U.S.S.R.	Heated part of reservoir	Pidgaiko et al. (1972)
Predator	6·6	Volchja R. res., U.S.S.R.	Heated part of reservoir	Pidgaiko et al. (1972)
Herbivore	41·0	Volchja R. res., U.S.S.R.	Unheated part of reservoir	Pidgaiko et al. (1972)
Predator	6·1	Volchja R. res., U.S.S.R.	Unheated part of reservoir	Pidgaiko et al. (1972)

Oligochaeta and Chironomids	15	Rybinsk res., U.S.S.R.		Sorokin (1972)
Predator	0·8	Rybinsk res., U.S.S.R.		Sorokin (1972)
Non-predator	19·2	8 lakes, excluding Kiev Reservoir	Ave. 8 lakes	Winberg (1972)
Non-predator	340·8	Kiev Reservoir		Winberg (1972)
Non-predator	18·1	3 lakes, U.S.S.R.	Ave. 3 lakes	Winberg et al. (1972)
Predator	5·6	3 lakes, U.S.S.R.	Ave. 3 lakes	Winberg et al. (1972)
Non-predator	4·7	L. Zelenetzkoye, U.S.S.R.	Subarctic lake	Winberg et al. (1973)
Predator	0·19	L. Zelenetzkoye, U.S.S.R.	Subarctic lake	Winberg et al. (1973)

to 10–12 years, where P/B ratios are generally much lower, down to 0·1 and 0·2 (Negus, 1966; Mann, K. H., 1971; Jónasson, 1972; Tudorancea, 1972; Burke and Mann, K. H., 1974). P/B ratios for tubificids and leeches (Hirudinea) appear more variable, perhaps reflecting difficulties in aging and growth rate determinations.

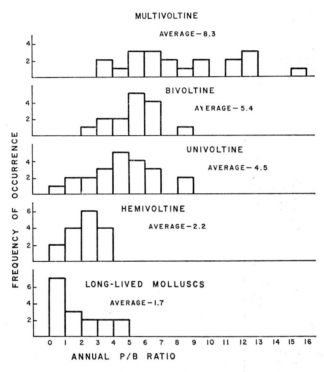

FIG. 4. Frequency occurrence of annual P/B ratios for zoobenthos reported in the literature.

Cohort P/B ratios, although only a few have been reported, show considerable constancy irrespective of the length of cohort life. These include: 2·9 and 3·4 for the leech *Erpobdella octoculata* (Elliott, 1973), 5·6–5·8 for the snail *Lymnaea palustris* (Hunter, 1975), 4·2 for the midge *Chironomus anthracinus* (Jónasson, 1972), 4·9 for the damselfly *Pyrrhosoma nymphula* (Lawton, 1971), 5·1 for the mayfly *Choroterpes mexicanus* (McClure and Stewart, 1976), 5·1–5·6 for the crayfish *Orconectes virilis* (Momot and Gowing, 1975, 1977), and 4·2 for the mayfly *Ephemerella subvaria* (Waters and Crawford, 1973), similar to the annual P/B ratios for univoltine species (Table IV).

C. ZOOPLANKTON

Annual production estimates for single-species zooplankters extend generally over a wider range than for zoobenthos, from near zero to usual maxima of about 30 g/m²/yr (dry weight) (= 300 kg/ha/yr) (Table VI, Fig. 5), including 28·7 g/m²/yr for species of the copepod *Epischura* in Lake Baikal (Moskalenko, 1971), 35·3 for *Daphnia hyalina* in a Welsh reservoir (George and Edwards, 1974) and 32·5 for *Daphnia schodleri* in Canyon Ferry reservoir, Montana (Wright, 1965). Topping all, however, was the estimate of 110 g/m²/yr for *Daphnia galeata mendotae* in Sanctuary Lake, Pennsylvania, reported by Cummins *et al.* (1969); this last estimate apparently was during a year and under circumstances that were particularly favorable for an extremely dense population of the daphnid; the following year production was only 7·4 g/m²/yr. These results illustrate, however, an extreme year-to-year variation that may be found in annual production and further emphasize the now widely-accepted tenet that broad ecological conclusions should not be made on temporally-limited observations.

Among the three groups included in the zooplankton, the rotifers showed the lowest single-species estimates and the cladocerans the highest, although estimates for both crustacean groups were similar except for a few outstandingly high estimates for the cladocerans. While not many estimates are available for the totals of the three groups considered separately, it appears that again the cladocerans as a group show the highest estimates, 20–25 g/m²/yr in two Polish lakes (Hillbricht-Ilkowska *et al.*, 1966). Estimates for the total of all zooplankton ranged higher, of course, up to about 70 g/m²/yr for the average of four Polish lakes, while one of these lakes produced nearly 100 (Kajak *et al.*, 1972). A few extremely low estimates for total zooplankton apparently resulted from the less favorable environmental conditions in cold, northern lakes (Winberg *et al.*, 1973).

Annual P/B ratios for single-species populations of zooplankton were much higher than for benthos because of the shorter life span and many more generations per year. Whereas no P/B ratios were reported for single species of rotifers, ratios available for total rotifer groups ranged usually from 40–75; exceptions are ratios of 8–10 for rotifers in the cold, northern lakes reported by Winberg *et al.* (1973). Among the annual P/B ratios reported for single-species copepods and cladocerans, while few, and totals for these groups as well, most are on the order of about 20. Lower P/B ratios probably can be attributed to estimates obtained for low-voltine copepods (Anderson, 1975; Bosselmann, 1975), predaceous plankters (Alimov and Winberg, 1972; Moskalenko and Votinsev, 1972), or in far northern or montane lakes

TABLE VI

Single-species and group estimates of annual production (g/m²—dry weight) and annual P/B ratios for zooplankton.
NA = data not available.

	P	P/B	Locality	Remarks	Authority
Rotifera					
Brachionus sp.	0·69	NA	Severson L., Minn.		Comita (1972)
Keratella quadrata	0·28	NA	Severson L., Minn.		Comita (1972)
Keratella cochlearis	0·068	NA	Severson L., Minn.		Comita (1972)
Polyarthra sp.	0·14	NA	Severson L., Minn.		Comita (1972)
Filinia longiseta	0·044	NA	Severson L., Minn.		Comita (1972)
Asplanchna sp.	0·82	NA	Severson L., Minn.		Comita (1972)
Synchaeta sp.	0·016	NA	Severson L., Minn.		Comita (1972)
Polyarthra vulgaris	0·04	NA	Lake Port-Bielh, France		Rey and Capblancq (1975)
Keratella cochlearis	0·003	NA	Char L., Cornwallis Is., Canada	Ave. 4 yrs, arctic lake	Rigler et al. (1974)
Total Rotifera					
	6·2	42	Taltowisko L., Poland	Mesotrophic lake	Hillbricht-Ilkowska et al. (1966)
	14·7	43	Mikolajskie L., Poland	Eutrophic lake	Hillbricht-Ilkowska et al. (1966)
	5·4	60	Rybinsk Reservoir, U.S.S.R.		Sorokin (1972)
	0·08	10·1	Lake Zelenetskoye, U.S.S.R.	Max. 2 yrs, cold north. lake	Winberg et al. (1973)
	0·05	8·0	Lake Akulkino, U.S.S.R.	Max. 2 yrs, cold north. lake	Winberg et al. (1973)
Non-predator	3·9	75	Three lakes, U.S.S.R.	Ave. 3 lakes	Winberg et al. (1972)
Predator	4·6	64	Three lakes, U.S.S.R.	Ave. 3 lakes	Winberg et al. (1972)

Copepoda

Species			Location	Notes	Reference
Eurycercus lamellatus	0·09	20	Lake Naroch, U.S.S.R.	Epiphytic copepod	Babitskiy (1970)
Eudiaptomus graciloides	7·7	6·8	Lake Esrom, Denmark	Bivoltine only	Bosselmann (1975)
Thermocyclops hyalinus	16·1	28·5	Lake George, Uganda		Burgis (1974)
Diaptomus siciloides	2·8	NA	Severson Lake, Minn.		Comita (1972)
Mesocyclops edax	0·74	NA	Severson Lake, Minn.		Comita (1972)
Cyclops vernalis	9·9	NA	Sanctuary Lake, Penn.	Max. 2 yrs	Cummins *et al.* (1969)
Diaptomus siciloides	9·4	NA	Sanctuary Lake, Penn.	Max. 2 yrs	Cummins *et al.* (1969)
Cyclops vicinus	10·2	21	Eglwys Nunydd res., S. Wales, U.K.		George (1976)
Acartia tonsa	17·1	NA	Patuxent R. estuary, Maryland	2 months, summer	Heinle (1966)
Diaptomus gracilis	10	NA	Thames Valley reservoirs, England	Ave. 2 reservoirs	Kibby (1971)
Cyclops vicinus	14·6	NA	Greifensee, Switz.		Mittelholzer (1970)
Eudiaptomus gracilis	7·8	NA	Greifensee, Switz.		Mittelholzer (1970)
Cyclops abyssorum	2·4	NA	Griefensee, Switz.		Mittelholzer (1970)
Mesocyclops leukarti	0·24	NA	Griefensee, Switz.		Mittelholzer (1970)
Eudiaptomus gracilis	1·9	NA	Lake of Lucerne, Switz.		Mittelholzer (1970)
Cyclops abyssorum	1·7	NA	Lake of Lucerne, Switz.		Mittelholzer (1970)
Cyclops vicinus	0·38	NA	Lake of Lucerne, Switz.		Mittelholzer (1970)
Mesocyclops leukarti	0·27	NA	Lake of Lucerne, Switz.		Mittelholzer (1970)
Epischura	28·7	NA	Lake Baikal, U.S.S.R.		Moskalenko (1971)
Mixodiaptomus laciniatus	1·3	2·5	Lake Port-Bielh, France	Ave. 6 yrs	Rey and Capblancq (1975)
Limnocalanus macrurus	0·29	NA	Char Lake, Cornwallis Is., Canada	Ave. 4 yrs, arctic lake	Rigler *et al.* (1974)
Total Copepoda	0·27	1·2	Snowflake L, Alberta	Univoltine, alpine lake	Anderson (1975)
	0·75	3·0	Teardrop Pond, Alberta	Bivoltine, alpine lake	Anderson (1975)

TABLE VI (*continued*)

	P	P/B	Locality	Remarks	Authority
	7·8	12	Taltowisko Lake, Poland	Mesotrophic lake	Hillbricht-Ilkowska et al. (1966)
	9·9	17·5	Mikolajskie Lake, Poland	Eutrophic lake	Hillbricht-Ilkowska et al. (1966)
	5·4	NA	Lake of Lucerne, Switz.		Mittelholzer (1970)
Cladocera					
Daphnia parvula	1·69	NA	Severson Lake, Minn.		Comita (1972)
Bosmina longirostris	1·49	NA	Severson Lake, Minn.		Comita (1972)
Diaphanosoma leuchtenbergianum	0·70	NA	Severson Lake, Minn.		Comita (1972)
Leptodora kindtii	2·1	NA	Sanctuary Lake, Penn.	Max. 2 yrs	Cummins et al. (1969)
Daphnia galeata mendotae	110	NA	Sanctuary Lake, Penn.	Max. 2 yrs	Cummins et al. (1969)
Ceriodaphnia reticulata	4·5	NA	Sanctuary Lake, Penn.	Max. 2 yrs	Cummins et al. (1969)
Bosmina longirostris	17·7	NA	Sanctuary Lake, Penn.	Max. 2 yrs	Cummins et al. (1969)
Chydorus sphaericus	4·0	NA	Sanctuary Lake, Penn.	Max. 2 yrs	Cummins et al. (1969)
Daphnia hyalina	35·3	23	Eglwys Nunydd res., S. Wales, U.K.	Ave. 2 yrs	George and Edwards (1974)
Diaphanosoma brachyurum	0·24	NA	Greifensee, Switz.		Mittelholzer (1970)
Daphnia longispina	0·56	2·5	Lake Port-Bielh, France		Rey and Capblancq (1975)
Daphnia longispina	8·7	NA	Lake Peter-Paul, Mich.	Lime-treated portion	Stross et al. (1961)
Daphnia longispina	7·2	NA	Lake Peter-Paul, Mich.	Untreated portion	Stross et al. (1961)
Daphnia schodleri	32·5	NA	Canyon Ferry res., Montana		Wright (1965)

			Location		Reference
Daphnia galeata mendotae	16·2	NA	Canyon Ferry res., Montana		Wright (1965)
Total Cladocera	19·2	22	Taltowisko Lake, Poland	Mesotrophic lake	Hillbricht-Ilkowska et al. (1966)
	24·7	31	Makolajskie Lake, Poland	Eutrophic lake	Hillbricht-Ilkowska et al. (1966)
Total Zooplankton					
Rotifera, Copepoda, Cladocera	1·98	14	Lake Krivoe, U.S.S.R.		Alimov et al. (1972)
Rotifera, Copepoda, Cladocera	1·60	13	Lake Krugloe, U.S.S.R.		Alimov et al. (1972)
Non-predator	3·4	13	Lake Krivoe, U.S.S.R.		Alimov and Winberg (1972)
Predator	0·52	5	Lake Krivoe, U.S.S.R.		Alimov and Winberg (1972)
Non-predator	11	13	Lake Krugloe, U.S.S.R.		Alimov and Winberg (1972)
Predator	2·0	11	Lake Krugloe, U.S.S.R.		Alimov and Winberg (1972)
Non-predator	17·7	18	Red Lake, U.S.S.R.		Andronikova et al. (1972)
Predator	2·0	15	Red Lake, U.S.S.R.		Andronikova et al. (1972)
Primary consumer	8·6	NA	Severson Lake, Minn.		Comita (1972)
Predator	0·83	NA	Severson Lake, Minn.		Comita (1972)
Rotifera, Copepoda, Cladocera	33·2	22	Taltowisko Lake, Poland	Mesotrophic lake	Hillbricht-Ilkowska et al. (1966)
Rotifera, Copepoda, Cladocera	49·3	27	Mikolajskie Lake, Poland	Eutrophic lake	Hillbricht-Ilkowska et al. (1966)
Non-predator	70·5	NA	Four lakes, Poland	Ave. 4 lakes	Kajak et al. (1972)

TABLE VI (*continued*)

	P	P/B	Locality	Remarks	Authority
Rotifera, Copepoda, Cladocera	23·3	NA	Taltowisko Lake, Poland	Mesotrophic lake	Kajak and Rybak (1966)
Rotifera, Copepoda, Cladocera	35	NA	Mikolajskie Lake, Poland	Eutrophic lake	Kajak and Rybak (1966)
Copepoda, Cladocera	48·7	NA	Greifensee, Switz.		Mittelholzer (1970)
Non-predator	16·1	13·5	Lake Baikal, U.S.S.R.		Moskalenko and Votinsev (1972)
Predator	1·6	2·4	Lake Baikal, U.S.S.R.		Moskalenko and Votinsev (1972)
Rotifera, Copepoda, Cladocera	1·9	NA	Lake Port-Bielh, France	3 Dominant spp.	Rey and Capblancq (1975)
Rotifera, Copepoda, Cladocera	8·8	NA	Rybinsk reservoir, U.S.S.R.		Sorokin (1972)
Non-predator	20·8	22	Lakes in U.S.S.R.	Ave. 9 lakes	Winberg (1972)
Rotifera, Crustacea	0·70	7·6	Lake Zelenetzkoye, U.S.S.R.	Max. 2 yrs, cold, north. lake	Winberg et al. (1973)
Rotifera, Crustacea	0·087	8·7	Lake Akulkino, U.S.S.R.	Max. 2 yrs, cold, north. lake	Winberg et al. (1973)

(Winberg *et al.*, 1973; Anderson, 1975; Rey and Capblancq, 1975). The voltinism of most plankton populations apparently was infrequently known, but except with some copepods which have long life spans, predaceous species and species in northern or mountain lakes, most crustacean species probably have four or more generations per year, and rotifers which mature and reproduce very rapidly likely have many more.

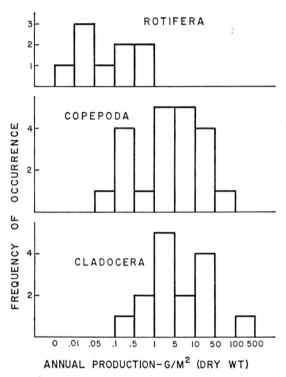

FIG. 5. Frequency occurrence of annual zooplankton production estimates reported in the literature. Note non-linear abscissal scale.

It would appear that most single-species estimates have resulted from North American or west European work, while estimates for totals of the three groups, and for total zooplankton as well, were provided in the Russian and east European literature. The reason for this is partly explained by the types of literature available to the writer while preparing this review: the papers most readily accessible from Russian and east European work were those summary- or review-type papers published in English (e.g. in the symposium edited by Kajak and Hillbricht-Ilkowska, 1972), or translations readily available.

Undoubtedly, a great quantity of data published by Russian and east European researchers include additional production estimates, particulary for zooplankton.

Another problem somewhat unique to zooplankton is that, whereas data are reported in this review on an annual basis, it is obvious that many annual estimates are only for the summer or "vegetative" season; probably the discrepancies are not great, as zooplankter populations exist, at least in important quantities in temperate zone lakes, only during the ice-free season. This is in contrast to zoobenthos, for which winter—even in temperate or northern lakes and streams—is a time of substantial production.

VI. Conclusions

Increasingly, measurement of production is being employed as an indicator of the health of an ecosystem, assessing the effect of environmental pollution or other disturbance. For example, Crisp, D. T. et al. (1974) measured fish production in an English stream system which would be inundated beneath a proposed impoundment, in order to compare with subsequent production data after impoundment; presumably, data will be forthcoming on production in the reservoir. Patalas (1970) and Pidgaiko et al. (1972) compared production of zooplankton and benthos in both heated and unheated portions of reservoirs used to cool power plants, as measures of possible detrimental effects from thermal effluents. Eckblad (1973) employed the production of gastropods in a New York reservoir as a measure of the effects of winter drawdown. In a series of production rate estimates on trout in Valley Creek, Minnesota, Elwood and Waters (1969) and Hanson and Waters (1974) documented the effect of floods on production, as well as recovery of production during the post-flood years. Similarly, Petrosky and Waters (1975) were able to evaluate the effect of an accidental turbidity and sedimentation problem in Valley Creek. In a Wisconsin trout stream, Brynildson and Mason (1975) ascertained the effect of pollution abatement with fish production data. Brylinsky and Mann, K. H. (1973) related secondary production to primary production to evaluate the effect of a number of environmental factors in lakes. In a quite different use of production estimates, Kimerle and Anderson (1971) estimated the production of a larval chironomid population as one means of energy transformation in a sewage lagoon in order to assess the lagoon's efficiency in sewage treatment.

Fish production has been used several times to evaluate sport fishery management programs for purposes of increasing yield to anglers, e.g. Hunt's (1969, 1971, 1974) extensive work in Lawrence

Creek, Wisconsin, which documented changes in trout production rate as the result of stream habitat alteration. Warren *et al.* (1964) evaluated the effect of artificial enrichment of a trout stream, as well as stream overhead canopy control, on fish production. And Stross *et al.* (1961) assessed zooplankton production as affected by liming in soft-water trout lakes.

With the exception of Hunt's work on Lawrence Creek, very little quantitative information has been obtained, relating total production to potential yield, either by sports anglers or in a commercial fishery. With his measures of the proportion of total fish production reaching the angler's creel, Hunt was able to document that one effect of the stream habitat management was to increase production in the larger sizes of fish, and thus also an increased yield of desirable-sized fish to the angler.

Considerable mention is made in the Russian literature of the importance on production data, for zooplankton and benthos as well as fish, to the commercial yield of fish from large east-European reservoirs. There appears little quantification of the relation between fish-food production and actual fish yield, at least in the literature available to the writer, but it is clear in many English summaries that the Russians place great significance on such lines of investigations.

Accumulation of empirically-based P/B ratios continues rapidly, although it is clear now that, except in unusual circumstances, the annual P/B ratio is a function of voltinism and may be assumed with fairly good precision, providing voltinism is known. The P/B ratio is not dependent on environmental factors such as fertility, temperature, etc. except indirectly as the environment may affect voltinism. As noted earlier, the annual P/B ratios suggested by Mann (1967) have proven accurate (cf. Fig. 4). Summarizing data from a number of lakes in the U.S.S.R. Winberg and Bauer (1971) concluded that annual P/B ratios in general for zooplankton were about 14–50 and for zoobenthos, 2·4–4·8.

One would not be too far off if the mean annual standing stock of the usual zooplankton population were multiplied by 15–20, or benthos by 6–8, or fish by 0·5–1 (1·2 for stream salmonids), to obtain an approximation of annual production. This technique is not of course a basic research tool; rather, it would be useful in a variety of applications, such as resource management. Increased use of the daily P/B ratio, or daily instantaneous growth rate (Zaika's (1973) "specific production") could well lead to much increased utility as an ecological research tool.

Improvement of methodology will no doubt continue. But probably more important at this time than conceptual models for production

F

calculation are improved field sampling techniques. The extreme sampling error involved, e.g. in stream bottom sampling, is so well recognized (and feared) that most researchers simply neglect it and forge ahead with no attempt at tests of statistical significance. In addition, systematic errors are also extremely serious, and even more insidious, for they frequently are not even detectable. For example, serious underestimates of growth rate due to such factors as continuous hatching of young (Macan, 1958) and size-differential predation on larger specimens (Ricker, 1969) are widely-known or suspected, but frequently the researcher has no means to detect them, much less measure them.

Representative sampling of the total population, both in space and through the whole life span, is another major problem. The discovery of the presence of a high proportion of stream benthic fauna deep in the hyporheic region of the substratum, has thrown considerable doubt upon the accuracy of previous stream zoobenthos production estimates and requires future researchers in this habitat to evaluate and measure this heretofore unsuspected part of the fauna (Coleman and Hynes, 1970; Williams and Hynes, 1974). The lack of concern for production in the very young and small stages of fishes has undoubtedly produced serious underestimates, as some researchers who have investigated this problem have shown (e.g. Mathews, 1971). Similarly, the production of gametes has been indicated by several workers to be a substantial, if not a major, proportion of annual fish production, as have also such losses as exuviae and secretions of animals such as case-building trichopterans.

Nevertheless, an analysis of production rates obtained for a wide variety of organisms, by a group of researchers perhaps of just as wide a variety, shows a rather remarkable degree of organization and order.

So, in summary, it appears that maximum production rates may easily reach levels of 1000–2000 kg/ha/yr (wet weight) for each of zooplankton, benthos and herbivorous-detritivorous fishes, probably several times these levels in the tropics, or where special areas of concentration occur, or in streams where the biota occupying a small aquatic area may profit from high rates of allochthonous import from a relatively large terrestrial area. Annual production rates for secondary consumers (predators) have been estimated at considerably lower rates; trophic level production efficiencies between trophic levels remain a fertile field for research, necessarily requiring quantities of production data that admittedly are expensive to accumulate, although some have already been roughed out. A fruitful goal of resource management would appear to be to improve these efficiencies, employing production data for evaluation.

ACKNOWLEDGEMENTS

I am especially grateful to Mrs Evelyn Gish, librarian of the Entomology, Fisheries, and Wildlife Library of the University of Minnesota,
for her generosity of time and gracious help in locating and acquiring
library materials. *The American Naturalist* gave permission for the
reproduction of the previously published Fig. 1. I am appreciative of
the stimulation I received from personal discussions long ago with
Richard O. Anderson and Edwin L. Cooper relative to concepts of
the P/B ratio. And I extend thanks for recent communications from
Andrew L. Hamilton and H. B. N. Hynes, who responded most helpfully
to my inquiries about certain methodological points in question.

Research grants by the National Science Foundation have been the
main source of support of studies in the University of Minnesota
Fisheries program concerned with freshwater production biology,
leading to several of the contributions quoted herein and ultimately
to the development of this review.

REFERENCES

Alexander, D. R. and MacCrimmon, H. R. (1974). Production and movement of
juvenile rainbow trout (*Salmo gairdneri*) in a headwater of Bothwell's Creek,
Georgian Bay, Canada. *J. Fish. Res. Bd. Can.* **31**, 117–121.
Alimov, A. F. and Winberg, G. G. (1972). Biological productivity of two northern
lakes. *Verh. Internat. Verein. Limnol.* **18**, 65-70.
Alimov, A. F., Boullion, V. V., Finogenova, N. P., Ivanova, M. B., Kuzmitskaya,
N. K., Nikulina, V. N., Ozertskovskaya, N. G. and Zharova, T. V. (1972).
Biological productivity of Lakes Krivoe and Krugloe. *In* "Productivity
Problems of Freshwaters" (Eds Z. Kajak and A. Hillbricht-Ilkowska).
IBP, UNESCO, Polish Sci. Publ., Warsaw, pp. 39–56.
Allen, K. R. (1949). Some aspects of the production and cropping of fresh waters.
Rpt. Sixth Sci. Congr. 1947. *Trans. R. Soc. N.Z.* **77** (5), 222–228.
Allen, K. R. (1950). The computation of production in fish populations. *N.Z. Sci.
Rev.* **8**, 89.
Allen, K. R. (1951). The Horokiwi Stream. *New Zealand Marine Dept., Fish.
Bull. No.* 10, 238 pp.
Allen, K. R. (1971). Relation between production and biomass. *J. Fish. Res. Bd.
Can.* **28**, 1573–1581.
Anderson, R. O. and Hooper, F. F. (1956). Seasonal abundance and production
of littoral bottom fauna in a southern Michigan lake. *Trans. Am. miscrosc.
Soc.* **75**, 259–270.
Anderson, R. S. (1975). An assessment of sport-fish production potential in two
small alpine waters in Alberta, Canada. *In* "Limnology of Shallow Waters"
(Eds J. Salanki and J. E. Ponyi). Symp. Biol. Hungary 15, Akad. Kiado,
Budapest, pp. 205–214.
Andersson, E. (1969). Life-cycle and growth of *Asellus aquaticus* (L.). *Rep. Inst.
Freshwat. Res. Drottningholm,* **49**, 5–26.
Andronikova, I. N., Drabkova, V. G., Kuzmenko, K. N., Michailova, N. F. and

152 THOMAS F. WATERS

Stravinskaya, E. A. (1972). Biological productivity of the main communities of the Red Lake. *In* "Productivity Problems of Freshwaters" (Eds Z. Kajak and A. Hillbricht-Ilkowska). IBP, UNESCO, Polish Sci. Publ., Warsaw, pp. 57-71.

Babitskiy, V. A. (1970). Biology and production of *Eurycercus lamellatus* (O. F. M.) along the shores of Lake Naroch. *Hydrobiol. J.* 6 (4), 26–32.

Backiel, T. (1967). An outline of methods for computation of fish production. *Ekol. Polsk. B.* 13, 197–211.

Backiel, T. (1971). Production and food consumption of predatory fish in the Vistula River. *J. Fish Biol.* 3, 369–405.

Balon, E. K. (1974). Fish production of a tropical ecosystem. *In* "Lake Kariba: A Man-made Tropical Ecosystem in Central Africa" (Eds E. K. Balon and A. G. Coche). W. Junk, Publ., The Hague, pp. 249–676.

Batzli, G. O. (1974). Production, assimilation and accumulation of organic matter in ecosystems. *J. theoret. Biol.* 45, 205–217.

Beattie, M., Bromley, H. J., Chambers, M., Goldspink, R., Vijverberg, J., van Zalinge, N. P. and Golterman, H. L. (1972). Limnological studies on Tjeukemeer—a typical Dutch "polder reservoir." *In* "Productivity Problems of Freshwaters" (Eds Z. Kajak and A. Hillbricht-Ilkowska). IBP, UNESCO, Polish Sci. Publ., Warsaw, pp. 421–446.

Benke, A. C. and Waide, J. B. (1977). In defense of average cohorts. *Freshwat. Biol.* 7, (in press).

Berrie, A. D. (1972). Productivity of the River Thames at Reading. *In* "Conservation and Productivity of Natural Waters" (Eds R. W. Edwards and D. J. Garrod). Symp. Zool. Soc. Lond. No. 29, Academic Press, London, pp. 69–86.

Biró, P. (1975). Observations on the fish production of Lake Balaton. *In* "Limnology of Shallow Waters" (Eds J. Salanki and J. E. Ponyi). Symp. Biol. Hungary 15, Akad. Kiado, Budapest, pp. 273–279.

Bishop, J. E. (1973). "Limnology of a Small Malayan River, Sungai Gombak." W. Junk, Publ., The Hague.

Borutsky, E. V. (1939). Dynamics of the total benthic biomass in the profundal of Lake Beloie. *Proc. Kossino Limnol. Sta., Hydrometerological Serv., U.S.S.R.* 22, 196–218. (In Russian, transl. by Michael Ovchynnyk.)

Bosselmann, S. (1975). Production of *Eudiaptomus graciloides* in Lake Esrom, 1970. *Arch. Hydrobiol.* 76, 43–64.

Boysen-Jensen, P. (1919). Valuation of the Limfjord. I. Studies on the fish-food in the Limfjord 1909–1917, its quantity, variation and annual production. *Report, Danish Biol. Sta., No.* 26, 3–44.

Brylinsky, M. and Mann, K. H. (1973). An analysis of factors governing productivity in lakes and reservoirs. *Limnol. Oceanogr.* 18, 1–14.

Brynildson, O. M. and Mason, J. W. (1975). Influence of organic pollution on the density and production of trout in a Wisconsin stream. *Wisc. Dept. Nat. Res., Tech. Bull. No.* 81, 15 pp.

Burgis, M. J. (1974). Revised estimates for the biomass and production of zooplankton in Lake George, Uganda. *Freshwat. Biol.* 4, 535–541.

Burke, M. V. and Mann, K. H. (1974). Productivity and production: biomass ratios of bivalve and gastropod populations in an eastern Canadian estuary. *J. Fish. Res. Bd. Can.* 31, 167–177.

Caspers, N. (1975a). Productivity and trophic structure of some West German woodland brooklets. *Verh. Internat. Verein. Limnol.* 19, 1712–1716.

Caspers, N. (1975b). Kalorische Werte der dominierenden Invertebraten zweir Waldbäche des Naturparkes Kottenforst-Ville. *Arch. Hydrobiol.* **75**, 484-489.

Cassie, R. M. (1950). The analysis of polymodal frequency distributions by the probability paper method. *N.Z. Sci. Rev.* **8**, 89–91.

Cassie, R. M. (1954). Some uses of probability paper in the analysis of size frequency distributions. *Aust. J. mar. Freshwat. Res.* **5**, 513–522.

Castro, L. B. (1975). Ökologie und Produktionsbiologie von *Agapetus fuscipes* Curt. im Breitenbach 1971–1972. Schlitzer Produktionsbiologische Studien (11). *Arch. Hydrobiol./Suppl.* **45**, 305–375.

Caswell, H. (1972). On instantaneous and finite birth rates. *Limnol. Oceanogr.* **17**, 787–791.

Chapman, D. W. (1965). Net production of juvenile coho salmon in three Oregon streams. *Trans. Am. Fish. Soc.* **94**, 40–52.

Chapman, D. W. (1967). Production in fish populations. *In* "The Biological Basis of Freshwater Fish Production" (Ed. S. D. Gerking). Wiley, New York, pp. 3–29.

Chapman, D. W. (1968). Production. *In* "Methods for Assessment of Fish Production in Fresh Waters" (Ed. W. E. Ricker). IBP Handbook No. 3, Blackwell Sci. Publ., Oxford and Edinburgh, pp. 182–196.

Chissenko, L. L. (1968). "Nomograms for Determining Weights of Water in Organisms from Dimensions and Form of the Body." Leningrad. pp. 1–166. (In Russian.)

Clarke, G. L. (1946). Dynamics of production in a marine area. *Ecol. Monogr.* **16**, 321–335.

Coche, A. G. (1967). Production of juvenile steelhead trout in a freshwater impoundment. *Ecol. Monogr.* **37**, 201–228.

Coffman, W. P., Cummins, K. W. and Wuycheck, J. C. (1971). Energy flow in a woodland stream ecosystem. I. Tissue support trophic structure of the autumnal community. *Arch. Hydrobiol.* **68**, 232–276.

Coleman, M. J. and Hynes, H. B. N. (1970). The vertical distribution of the invertebrate fauna in the bed of a stream. *Limnol. Oceanogr.* **15**, 31–40.

Comita, G. W. (1972). The seasonal zooplankton cycles, production and transformations of energy in Severson Lake, Minnesota. *Arch. Hydrobiol.* **70**, 14–66.

Cook, D. G. and Johnson, M. G. (1974). Benthic macroinvertebrates of the St. Lawrence Great Lakes. *J. Fish. Res. Bd. Can.* **31**, 763–782.

Cooper, E. L. and Scherer, R. C. (1967). Annual production of brook trout (*Salvelinus fontinalis*) in fertile and infertile streams of Pennsylvania. *Proc. Pa. Acad. Sci.* **41**, 65–70.

Cooper, W. E. (1965). Dynamics and production of a natural population of a fresh-water amphipod, *Hyalella azteca*. *Ecol. Monogr.* **35**, 377–394.

Crisp, D. J. (1971). Energy flow measurements. *In* "Methods for the Study of Marine Benthos" (Eds N. A. Holme and A. D. McIntyre). IBP Handbook No. 16, Blackwell Sci. Publ., Oxford and Edinburgh, pp. 197–279.

Crisp, D. J. (1975). Secondary productivity in the sea. *In* "Productivity of World Ecosystems". Natl. Acad. Sci., Washington, pp. 71–89.

Crisp, D. T. (1962). Estimates of the annual production of *Corixa germari* (Fieb.) in an upland reservoir. *Arch. Hydrobiol.* **58**, 210–223.

Crisp, D. T., Mann, R. H. K. and McCormack, J. C. (1974). The populations of fish at Cow Green, Upper Teesdale, before impoundment. *J. appl. Ecol.* **11**, 969–996.

Crisp, D. T., Mann, R. H. K. and McCormack, J. C. (1975). The populations of fish in the River Tees system on the Moor House National Nature Reserve, Westmorland. *J. Fish Biol.* **7**, 573–593.

Cummins, K. W. and Wuycheck, J. C. (1971). Caloric equivalents for investigations in ecological energetics. *Mitt. Internat. Verein. Limnol.*, No. 18, 158 pp.

Cummins, K. W., Costa, R. R., Rowe, R. E., Moshiri, G. A., Scanlon, R. M. and Zajdel, R. K. (1969). Ecological energetics of a natural population of the predaceous zooplankter *Leptodora kindtii* Focke (Cladocera). *Oikos* **20**, 189–223.

Cushman, R. M., Elwood, J. W. and Hildebrand, S. G. (1975). Production dynamics of *Alloperla mediana* Banks (Plecoptera: Chloroperlidae) and *Diplectrona modesta* Banks (Trichoptera: Hydropsychidae) in Walker Branch, Tennessee. *Oak Ridge Nat. Lab., Environ. Sci. Div. Publ. No.* 785, 66 pp.

Décamps, H. and Lafont, M. (1974). Cycles vitaux et production des *Micrasema* Pyreneennes dans les mousses d'eau courante (Trichoptera, Brachycentridae). *Annls. Limnol.* **10**, 1–32.

Driver, E. A., Sugden, L. G. and Kovach, R. J. (1974). Calorific, chemical and physical values of potential duck foods. *Freshwat. Biol.* **4**, 281–292.

Dumont, H. J. (1975). The dry weight estimate of biomass in a selection of Cladocera, Copepoda and Rotifera from the plankton, periphyton and benthos of continental waters. *Oecologia* **19**, 75–97.

Duncan, A. (1975). Production and biomass of three species of *Daphnia* coexisting in London reservoirs. *Verh. Internat. Verein. Limnol.* **19**, 2858–2867.

Eckblad, J. W. (1973). Population studies of three aquatic gastropods in an intermittent backwater. *Hydrobiologia* **41**, 199–219.

Edmondson, W. T. (1946). Factors in the dynamics of rotifer populations. *Ecol. Monogr.* **16**, 357-372.

Edmondson, W. T. (1960). Reproductive rates of rotifers in natural populations. *Mem. Ist. Ital. Idrobiol.* **12**, 21–77.

Edmondson, W. T. (1968). A graphical method for evaluating the use of the egg ratio for measuring birth and death rates. *Oecologia* **1**, 1–37.

Edmondson, W. T. (1971). Reproductive rate determined indirectly from egg ratio. *In* "A Manual on Methods for the Assessment of Secondary Productivity in Fresh Waters" (Eds W. T. Edmondson and G. G. Winberg). IBP Handbook No. 17, Blackwell Sci. Publ., Oxford and Edinburgh, pp. 165–169.

Edmondson, W. T. (1972). Instantaneous birth rates of zooplankton. *Limnol. Oceanogr.* **17**, 792–795.

Edmondson, W. T. (1974). Secondary production. *Mitt. Internat. Verein. Limnol.* **20**, 229–272.

Edmondson, W. T. and Winberg, G. G. (Eds) (1971). "A Manual on Methods for the Assessment of Secondary Productivity in Fresh Waters." IBP Handbook No. 17, Blackwell Sci. Publ., Oxford and Edinburgh.

Edwards, R. W. and Garrod, D. J. (Eds) (1972). "Conservation and Productivity of Natural Waters." Symp. Zool. Soc. Lond. No. 29, Academic Press, London.

Egglishaw, H. J. (1970). Production of salmon and trout in a stream in Scotland. *J. Fish Biol* **2**, 117–136.

Elliott, J. M. (1973). The life cycle and production of the leech *Erpobdella octoculata* (L.) (Hirudinea: Erpobdellidae) in a Lake District stream. *J. Anim. Ecol.* **42**, 435–448.

Elster, H. I. (1954). Über die Populationsdynamik von *Eudiaptomus gracilis*

Sars und *Heterocope borealis* Fischer im Bodensee-Obersee. *Arch. Hydrobiol./ Suppl.* **20**, 546–614.

Elwood, J. W. and Waters, T. F. (1969). Effects of floods on food consumption and production rates of a stream brook trout population. *Trans. Am. Fish. Soc.* **98**, 253–262.

Erman, D. C. and Erman, N. A. (1975). Macroinvertebrate composition and production in some Sierra Nevada minerotrophic peatlands. *Ecology* **56**, 591–603.

Fager, E. W. (1969). Production of stream benthos: A critique of the method of assessment proposed by Hynes and Coleman (1968). *Limnol. Oceanogr.* **14**, 766–770.

Fisher, S. G. and Likens, G. E. (1973). Energy flow in Bear Brook, New Hampshire: An integrative approach to stream ecosystem metabolism. *Ecol. Monogr.* **43**, 421–439.

Fuller, W. A. and Kevan, P. G. (Eds) (1970). "Productivity and Conservation in Northern Cicumpolar Lands." Proc., Conf. IBP, Edmonton, Alberta, Canada, 1969. Internat. Union for Cons. of Nature and Nat'l. Resources, Morges, Switz.

Gehrs, C. W. and Robertson, A. (1975). Use of life tables in analyzing the dynamics of copepod populations. *Ecology* **56**, 665–672.

Geiling, W. T. and Campbell, R. S. (1972). The effect of temperature on the development rate of the major life stages of *Diaptomus pallidus* Herrick. *Limnol. Oceanogr.* **17**, 304–307.

George, D. G. (1976). Life cycle and production of *Cyclops vicinus* in a shallow eutrophic reservoir. *Oikos* **27**, 101–110.

George, D. G. and Edwards, R. W. (1974). Population dynamics and production of *Daphnia hyalina* in a eutrophic reservoir. *Freshwat. Biol.* **4**, 445—465.

Gerking, S. D. (1962). Production and food utilization in a population of bluegill sunfish. *Ecol. Monogr.* **32**, 31–78.

Gerking, S. D. (Ed.) (1967). "The Biological Basis of Freshwater Fish Production." IBP Symp., Productivity of Freshwater Communities, Reading, 1966. Wiley, New York.

Giani, N. and Laville, H. (1973). Cycle biologique et production de *Sialis lutaria* L. (Megaloptera) dans le Lac de Port-Bielh (Pyrénées Centrales). *Annls. Limnol.* **9**, 45–61.

Gillespie, D. M. (1969). Population studies of four species of molluscs in the Madison River, Yellowstone National Park. *Limnol. Oceanogr,* **14**, 101–114.

Goldman, Charles R. (Ed.) (1966). "Primary Productivity in Aquatic Environments." IBP Symp., Pallanza, Italy, 1965. Univ. Calif. Press, Berkeley and Los Angeles.

Golley, F. B. (1961). Energy values of ecological materials. *Ecology* **42**, 581–584.

Golley, F. B. and Buechner, H. K. (Eds) (1968). "A Practical Guide to the Study of the Productivity of Large Herbivores." IBP Handbook No. 7, Blackwell Sci. Publ., Oxford and Edinburgh.

Goodnight, W. H. and Bjornn, T. C. (1971). Fish production in two Idaho streams. *Trans. Am. Fish. Soc.* **100**, 769–780.

Greze, V. N. (1965). Growth rate and production potential of fish populations. *Gidrobiol. Zh.* **1** (2), 35–42. (In Russian, Transl. Series 897, Fish. Res. Bd. Can., 1967).

Hall, D. J. (1964). An experimental approach to the dynamics of a natural population of *Daphnia galeata mendotae. Ecology* **45**, 94–112.

Hall, D. J., Cooper, W. E. and Werner, E. E. (1970). An experimental approach to the production dynamics and structure of freshwater animal communities. *Limnol. Oceanogr.* **15**, 839–928.

Hall, K. J. and Hyatt, K. D. (1974). Marion Lake (IBP)—From bacteria to fish. *J. Fish. Res. Bd. Can.* **31**, 893–911.

Hamilton, A. L. (1969). On estimating annual production. *Limnol. Oceanogr.* **14**, 771–782.

Hanson, D. L. and Waters, T. F. (1974). Recovery of standing crop and production rate of a brook trout population in a flood-damaged stream. *Trans. Am. Fish. Soc.* **103**, 431–439.

Harding, J. P. (1949). The use of probability paper for the graphical analysis of polymodal frequency distributions. *J. mar. biol. Ass. U.K.* **28**, 141–153.

Hart, R. C. and Allanson, B. R. (1975). Preliminary estimates of production by a calanoid copepod in subtropical Lake Sibaya. *Verh. Internat. Verein. Limnol.* **19**, 1434–1441.

Hatch, R. W. and Webster, D. A. (1961). Trout production in four central Adirondack Mountain lakes. *Cornell. Univ Mem.* **373**, 82 pp.

Hayne, D. W. and Ball, R. C. (1956). Benthic productivity as influenced by fish predation. *Limnol. Oceanogr.* **1**, 161–175.

Heinle, D. R. (1966). Production of a calanoid copepod, *Acartia tonsa*, in the Patuxent River estuary. *Chesapeake Sci.* **7**, 59–74.

Hillbricht-Ilkowska, A., Gliwicz, Z. and Spodniewska, I. (1966). Zooplankton production and some trophic dependences in the pelagic zone of two Masurian lakes. *Verh. Internat. Verein. Limnol.* **16**, 432–440.

Holcik, J. (1972). Abundance, ichthyomass and production of fish populations in three types of water-bodies in Czechoslovakia (man-made lake, trout lake, arm of the Danube River). *In* "Productivity Problems of Freshwaters" (Eds Z. Kajak and A. Hillbricht-Ilkowska). IBP, UNESCO, Polish Sci. Publ. Warsaw, pp. 843–855.

Holcik, J. and Pivnicka, K. (1972). The density and production of fish populations in the Klicava Reservoir (Czechoslovakia) and their changes during the period 1957–1970. *Int. Revue ges. Hydrobiol.* **57**, 883–894.

Holme, N. A. and McIntyre, A. D. (Eds) (1971). "Methods for the Study of Marine Benthos." IBP Handbook No. 16, Blackwell Sci. Publ., Oxford and Edinburgh.

Hopkins, C. L. (1971). Production of fish in two small streams in the North Island of New Zealand. *N.Z. J. Mar. Freshwat. Res.* **5**, 280–290.

Horst, T. J. and Marzolf, G. R. (1975). Production ecology of burrowing mayflies in a Kansas reservoir. *Verh. Internat. Verein. Limnol.* **19**, 3029–3038.

Horton, P. A. (1961). The bionomics of brown trout in a Dartmoor stream. *J. Anim. Ecol.* **30**, 311–338.

Horton, P. A., Baily, R. G. and Wilsdon, S. I. (1968). A comparative study of the bionomics of the salmonids of three Devon streams. *Arch. Hydrobiol.* **65**, 187–204.

Hudson, P. L. and Swanson, G. A. (1972). Production and standing crop of *Hexagenia* (Ephemeroptera) in a large reservoir. *Studies in Nat. Sci., Nat. Sci. Res. Inst., Eastern N. M. Univ.*, Vol. 1, No. 4, 42 pp.

Huet, M. (1964). The evaluation of the fish productivity in fresh waters. *Verh. Internat. Verein. Limnol.* **15**, 524–528.

Hunt, R. L. (1966). Production and angler harvest of wild brook trout in Lawrence Creek, Wisconsin. *Wisc. Cons. Dept., Tech. Bull. No.* 35, 52 pp.

Hunt, R. L. (1969). Effects of habitat alteration on production, standing crops and yield of brook trout in Lawrence Creek, Wisconsin. *In* "Symposium on Salmon and Trout in Streams" (Ed. T. G. Northcote). H. R. MacMillan Lectures in Fisheries, Univ. British Columbia, Vancouver, pp. 281–312.

Hunt, R. L. (1971). Responses of a brook trout population to habitat development in Lawrence Creek. *Wisc. Dept. Nat. Res., Tech., Bull. No.* 48, 35 pp.

Hunt, R. L. (1974). Annual production by brook trout in Lawrence Creek during eleven successive years. *Wisc. Dept. Nat. Res., Tech. Bull. No.* 82, 29 pp.

Hunter, R. D. (1975). Growth, fecundity, and bioenergetics in three populations of *Lymnaea palustris* in upstate New York. *Ecology* **56**, 50–63.

Hynes, H. B. N. (1961). The invertebrate fauna of a Welsh mountain stream. *Arch. Hydrobiol.* **57**, 344–388.

Hynes, H. B. N. (1970). "The Ecology of Running Waters." Univ. Toronto Press, Toronto.

Hynes, H. B. N. and Coleman, M. J. (1968). A simple method of assessing the annual production of stream benthos. *Limnol. Oceanogr.* **13**, 569–573.

Illies, J. (1975). A new attempt to estimate production in running waters. (Schlitz studies on productivity, No. 12). *Verh. Internat. Verein. Limnol.* **19**, 1705–1711.

Ivanova, M. B. (1973). Some remarks on the method of production estimation. *Polsk. Arch. Hydrobiol.* **20**, 435–441.

Ivlev, V. S. (1945). The biological productivity of waters. (English version, 1966, *J. Fish. Res. Bd. Can.* **23**, 1727–1759.)

Johnson, M. G. (1974). Production and productivity. *In* "The Benthos of Lakes" (Ed. R. O. Brinkhurst). St. Martin's Press, New York, pp. 46–64.

Johnson, M. G. and Brinkhurst, R. O. (1971). Production of benthic macroinvertebrates of Bay of Quinte and Lake Ontario. *J. Fish. Res. Bd. Can.* **28**, 1699–1714.

Jónasson, P. M. (1972). Ecology and production of the profundal benthos in relation to phytoplankton in Lake Esrom. *Oikos (Suppl.)* **14,**, 1–148.

Jónasson, P. M. (1975). Population ecology and production of benthic detritivores. *Verh. Internat. Verein. Limnol.* **19**, 1066–1072.

Juday, C. (1940). The annual energy budget of an inland lake. *Ecology* **21**, 438–450.

Kajak, Z. (1967). Remarks on methods of investigating benthos production. *Ekol. Polsk. B.* **13**, 173–195.

Kajak, Z. and Hillbricht-Ilkowska, A. (Eds) (1972). "Productivity Problems of Freshwaters." IBP, UNESCO, Polish Sci. Publ., Warsaw.

Kajak, Z. and Rybak, J. I. (1966). Production and some trophic dependences in benthos against primary production and zooplankton production of several Masurian lakes. *Verh. Internat. Verein Limnol.* **16**, 441–451.

Kajak, Z., Hillbricht-Ilkowska, A. and Pieczynska, E. (1972). The production processes in several Polish lakes. *In* "Productivity Problems of Freshwaters," (Eds Z. Kajak and A. Hillbricht-Ilkowska). IBP, UNESCO, Polish Sci. Publ., Warsaw, pp. 129–147.

Kay, D. G. and Brafield, A. E. (1973). The energy relations of the polychaete *Neanthes* (= *Nereis*) *virens* (Sars). *J. Anim. Ecol.* **42**, 673–692.

Kelso, J. R. M. and Ward, F. J. (1972). Vital statistics, biomass, and seasonal production of an unexploited walleye (*Stizostedion vitreum vitreum*) population in West Blue Lake, Manitoba. *J. Fish. Res. Bd. Can.* **29**, 1043–1052.

Kibby, H. V. (1971). Energetics and population dynamics of *Diaptomus gracilis*. *Ecol. Monogr.* **41**, 311–327.

Kimerle, R. A. and Anderson, N. H. (1971). Production and bioenergetic role of the midge *Glyptotendipes barbipes* (Staeger) in a waste stabilization lagoon. *Limnol. Oceanogr.* **16**, 646–659.

Kipling, C. and Frost, W. E. (1970). A study of the mortality, population numbers, year class strengths, production and food consumption of pike, *Esox lucius* L., in Windermere from 1944 to 1962. *J. Anim. Ecol.* **39**, 115–157.

Konstantinov, A. S. and Nechvalenko, S. P. (1968). On the accuracy of determining the production of chironomids by the method of summing their daily growth increments. *Gidrobiol. Zh.* **4** (6): 77–82. (In Russian, transl. by W. E. Ricker, *Fish. Res. Bd. Can. Transl. Ser. No.* 1368, 1970.)

Konstantinova, N. S. (1961). On the growth of Cladocera and measurement of their production. *Veprosy Ikhtiologii Vol.* 1, *No.* 2 (19): 363–367. (In Russian, transl. by R. E. Foerster, *Fish. Res. Bd. Can. Transl. Ser. No.* 410, 1962).

Korimek, V. (1966). The production of adult females of *Daphnia pulicaria* Forbes in a carp pond estimated by a direct method. *Verh. Internat. Verein. Limnol.* **16**, 386–391.

Kuz'menko, K. N. (1969). The life cycle and production of *Pontoporeia affinis* Lindstr. in Lake Krasnoye (Karelian isthmus). *Hydrobiol. J.* **5** (6A), 40–45. (In Russian, transl. pub. by Am. Fisheries Soc.)

Ladle, M., Bass, J. A. B. and Jenkins, W. R. (1972). Studies on production and food consumption by the larval Simuliidae (Diptera) of a chalk stream. *Hydrobiologia* **39**, 429–448.

Lair, N. (1975). Sur la production des Copepodes dans deux lacs du Massif Central français. *Verh. Internat. Verein. Limnol.* **19**, 3204–3211.

LaRow, E. J. (1975). Secondary productivity of *Leptodora kindtii* in Lake George, N.Y. *Am. Midl. Nat.* **94**, 120–126.

Lavandier, P. (1975). Cycle biologique et production de *Capnioneura brachyptera* D. (Plecopteres) dans un ruisseau d'altitude des Pyrénées centrales. *Annls. Limnol.* **11**, 145–156.

Laville, H. (1971). Recherches sur les Chironomides (Diptera) lacustres du Massif de Neouvielle (Hautes-Pyrénées). Deux. Part. Communautes et Production. *Annls. Limnol.* **7**, 335–414.

Laville, H. (1975). Production d'un Chironomide semivoltin (*Chironomus commutatus* Str.) dans le Lac de Port-Bielh (Pyrénées Centrales). *Annls. Limnol.* **11**, 67–77.

Lawton, J. H. (1971). Ecological energetics studies on larvae of the damselfly *Pyrrhosoma nymphula* (Sulzer) (Odonata: Zygoptera). *J. Anim. Ecol.* **40**, 385–423.

Learner, M. A. and Potter, D. W. B. (1974). Life-history and production of the leech *Helobdella stagnalis* (L.) (Hirudinea) in a shallow eutrophic reservoir in South Wales. *J. Anim. Ecol.* **43**, 199–208.

Le Cren, E. D. (1962). The efficiency of reproduction and recruitment in freshwater fish. *In* "The Exploitation of Natural Animal Populations" (Eds E. D. Le Cren and M. W. Holdgate). Symp. British Ecol. Soc., No. 2, Wiley, New York, pp. 283–296.

Le Cren, E. D. (1969). Estimates of fish populations and production in small streams in England. *In* "Symposium on Salmon and Trout in Streams" (Ed. T. G. Northcote). H. R. MacMillan Lectures in Fisheries, Univ. British Columbia, Vancouver, pp. 269–280.

Le Cren, E. D. (1972). Fish production in freshwaters. *In* "Conservation and Productivity of Natural Waters" (Eds R. W. Edwards and D. J. Garrod). Symp. Zool. Soc. London, Academic Press, London, pp. 115–133.

Leveque, C. (1973). Dynamique des peuplements biologie, et estimation de la production des mollusques benthique du Lac Tchad. *Cah. O.R.S.T.O.M., Ser. Hydrobiol.* **7**, 117–147.

Lindeman, R. L. (1941). Seasonal food-cycle dynamics in a senescent lake. *Am. Midl. Nat.* **26**, 636–673.

Lotrich, V. A. (1973). Growth, production, and community composition of fishes inhabiting a first-, second-, and third-order stream of eastern Kentucky. *Ecol. Monogr.* **43**, 377–397.

Lowry, G. R. (1966). Production and food of cutthroat trout in three Oregon coastal streams. *J. Wildl. Mgmt,* **30**, 754–767.

Lundbeck, J. (1926). Die Bodentierweld Norddeutscher Seen. *Arch. Hydrobiol./ Suppl.* **7**, 1–473.

Macan, T. T. (1958). Causes and effects of short emergence periods in insects. *Verh. Internat. Verein. Limnol.* **13**, 845–849.

Maitland, P. S. and Hudspith, P. M. G. (1974). The zoobenthos of Loch Leven, Kinross, and estimates of its production in the sandy littoral area during 1970 and 1971. *Proc. R. Soc. Edinb. B.,* **74**, 219–239.

Maitland, P. S., Charles, N. W., Morgan, N. C., East, K. and Gray, M. C. (1972). Preliminary research on the production of Chironomidae in Loch Leven, Scotland. *In* "Productivity Problems of Freshwaters" (Eds Z. Kajak and A. Hillbricht-Ilkowska). IBP, UNESCO, Polish Sci. Publ., Warsaw, pp. 795–812.

Mann, K. H. (1964). The pattern of energy flow in the fish and invertebrate fauna of the River Thames. *Verh. Internat. Verein. Limnol.* **15**, 485–495.

Mann, K. H. (1965). Energy transformations by a population of fish in the River Thames. *J. Anim. Ecol.* **34**, 253–275.

Mann, K. H. (1967). The cropping of the food supply. *In* "The Biological Basis of Freshwater Fish Production" (Ed. S. D. Gerking). IBP Symp., Productivity of Freshwater Communities, Reading, 1966. Wiley, New York, pp. 243–257.

Mann, K. H. (1969). The dynamics of aquatic ecosystems. *Adv. Ecol. Research* **6**, 1–81.

Mann, K. H. (1971). Use of the Allen curve method for calculating benthic production. *In* "A Manual on Methods for the Assessment of Secondary Productivity in Fresh Waters" (Eds W. T. Edmondson and G. G. Winberg). IBP Handbook No. 17, Blackwell Sci. Publ., Oxford and Edinburgh, pp. 160–165.

Mann, K. H., Britton, R. H., Kowalczewski, A., Lack, T. J., Mathews, C. P. and McDonald, I. (1972). Productivity and energy flow at all tophic levels in the River Thames, England. *In* "Productivity Problems of Freshwaters" (Eds Z. Kajak and A. Hillbricht-Ilkowska). IBP, UNESCO, Polish Sci. Publ., Warsaw, pp. 579–596.

Mann, R. H. K. (1971). The populations, growth and production of fish in four small streams in southern England. *J. Anim. Ecol.* **40**, 155–190.

Mathews, C. P. (1970). Estimates of production with reference to general surveys. *Oikos* **21**, 129–133.

Mathews, C. P. (1971). Contribution of young fish to total production of fish in the River Thames near Reading. *J. Fish Biol.* **3**, 157–180.

Mathews, C. P. and Westlake, D. F. (1969). Estimation of production by populations of higher plants subject to high mortality. *Oikos* **20**, 156–160.

Mathias, J. A. (1971). Energy flow and secondary production of the amphipods *Hyalella azteca* and *Crangonyx richmondensis occidentalis* in Marion Lake, British Columbia. *J. Fish. Res. Bd. Can.* **28**, 711–726.

McClure, R. G. and Stewart, K. W. (1976). Life cycle and production of the mayfly *Choroterpes* (*Neochoroterpes*) *mexicanus* Allen (Ephemeroptera: Leptophlebiidae. *Annls. Ent. Soc. Am.* **69**, 134–144.

Miller, R. B. (1941). A contribution to the ecology of the Chironomidae of Costello Lake, Algonquin Park, Ontario. *Publ. Ont. Fish. Res. Lab., No.* 60. *Univ. Toronto Studies, Biol. Series* **49**, 1–63.

Mills, D. H. (1967). A study of trout and young salmon populations in forest streams with a view to management. *Forestry* (Oxford, England) **40** (1), *Suppl.*, 85–90.

Mittelholzer, E. (1970). Populationsdynamik und Produktion des Zooplanktons im Greifensee und in Vierwaldstättersee. *Schweiz. Z. Hydrol.* **32**, 90–149.

Momot, W. T. (1967). Population dynamics and productivity of the crayfish, *Orconectes virilis*, in a marl lake. *Am. Midl. Nat.* **78**, 55–81.

Momot, W. T. and Gowing, H. (1975). The cohort production and life cycle turnover ratio of the crayfish, *Orconectes virilis*, in three Michigan lakes. *In* "Freshwater Crayfish" (Ed. J. W. Avault). 2nd Internat. Crayfish Symp., 1974. Louisiana St. Univ., Baton Rouge, pp. 489–511.

Momot, W. T. and Gowing, H. (1977). Response of the crayfish, *Orconectes virilis*, to experimental fishing. *J. Fish. Res. Bd. Can.* (In Press).

Mori, S. and Yamamoto, G. (Eds) (1975). "Productivity of Communities in Japanese Inland Waters." Jap. Comm. for IBP, Univ. Tokyo Press, Tokyo.

Moshiri, G. A. and Cummins, K. W. (1969). Calorific values for *Leptodora kindtii* Focke (Crustacea Cladocera) and selected food organisms. *Arch. Hydrobiol.* **66**, 91–99.

Moskalenko, B. K. (1971). The biological productivity of Lake Baykal. *Hydrobiol. J.* **7** (5), 1–8.

Moskalenko, B. K. and Votinsev, K. K. (1972). Biological productivity and balance of organic substance and energy in Lake Baikal. *In* "Productivity Problems of Freshwaters" (Eds Z. Kajak and A. Hillbricht-Ilkowska). IBP, UNESCO, Polish Sci. Publ., Warsaw, pp. 207–226.

Mullin, M. M. (1969). Production of zooplankton in the ocean: The present status and problems. *Oceanogr. Mar. Biol., Ann. Rev.* **7**, 293–314.

National Academy of Sciences. (1975). "Productivity of World Ecosystems." U.S. Nat. Comm., IBP, Washington.

Neess, J. and Dugdale, R. C. (1959). Computation of production for populations of aquatic midge larvae. *Ecology* **40**, 425–430.

Negus, C. (1966). A quantitative study of growth and production of Unionid mussels in the River Thames at Reading. *J. Anim. Ecol.* **35**, 513–532.

Nelson, D. J. and Scott, D. C. (1962). Role of detritus in the productivity of a rock-outcrop community in a Piedmont stream. *Limnol. Oceanogr.* **7**, 396–413.

Neveu, A. (1973). Estimation de la production de populations larvaires du genre *Simulium* (Diptera, Nematocera). *Ann. Hydrobiol.* **4**, 183–199.

O'Connor, J. F. and Power, G. (1973). Trout production and eels in Bill Lake, Saguenay County, Quebec. *J. Fish. Res. Bd. Can.* **30**, 1398–1401.

O'Connor, J. F. and Power, G. (1976). Production by brook trout (*Salvelinus fontinalis*) in four streams in the Matamek watershed, Quebec. *J. Fish. Res. Bd. Can.* **33**, 6–18.

Odum, E. P. (1959). "Fundamentals of Ecology", 2nd ed. Saunders, Philadelphia and London.

Odum, H. T. (1957). Trophic structure and productivity of Silver Springs, Florida. *Ecol. Monogr.* **27**, 55–112.

Otto, C. (1975). Energetic relationships of the larval population of *Potamophylax cingulatus* (Trichoptera) in a South Swedish stream. *Oikos* **26**, 157–169.

Paine, R. T. (1971). The measurement and application of the calorie to ecological problems. *Ann. Rev. Ecol., Syst.* **2**, 145–164.

Paloheimo, J. E. (1974). Calculation of instantaneous birth rate. *Limnol. Oceanogr.* **19**, 692–694.

Patalas, K. (1970). Primary and secondary production in a lake heated by thermal power plant. *Proc. Inst. Environ. Sci.*, 16th Ann. Tech. Meeting, Boston, pp. 267–271.

Pearson, W. D. and Kramer, R. H. (1972). Drift and production of two aquatic insects in a mountain stream. *Ecol. Monogr.* **24**, 365–385.

Pechen, G. A. and Shushkina, E. A. (1964). The production of planktonic crustaceans in lakes of diverse types. *Biol. osnov. rybn. kh-va na vnutr. vodoemakh Pribaltiki.* Minsk, pp. 249–257. (In Russian, quoted by Winberg, 1971.)

Petrosky, C. E. and Waters, T. F. (1975). Annual production by the slimy sculpin population in a small Minnesota trout stream. *Trans. Am. Fish. Soc.* **104**, 237–244.

Petrusewicz, K. (Ed.) (1967). "Secondary Productivity of Terrestrial Ecosystems." Proc. Working Meeting, IBP, Jablonna, Poland, 1966. Polish Sci. Publ., Warsaw.

Petrusewicz, K. and Macfadyen, A. (1970). "Productivity of Terrestrial Animals, Principles and Methods." IBP Handbook No. 13. F. A. Davis Co., Philadelphia.

Petrusewicz, K. and Ryszkowski, L. (Eds) (1969/1970). "Energy flow through Small Mammal Populations." Proc. IBP Meeting on Secondary Productivity in Small Mammal Populations, Oxford, 1968. Polish Sci. Publ., Warsaw.

Pidgaiko, M. L., Grin, V. G., Kititsina, L. A., Lenchina, L. G., Polivannaya, M. F., Sergeeva, O. A. and Vinogradskaya, T. A. (1972). Biological productivity of Kurakhov's Power Station cooling reservoir. *In* "Productivity Problems of Freshwaters" (Eds Z. Kajak and A. Hillbricht-Ilkowska). IBP, UNESCO, Polish Sci. Publ., Warsaw, pp. 477–491.

Platt, T., Brawn, V. M. and Irwin, B. (1969). Carbon and calorie equivalents of zooplankton biomass. *J. Fish. Res. Bd. Can.* **26**, 2345–2349.

Potter, D. W. B. and Learner, M. A. (1974). A study of the benthic macro-invertebrates of a shallow eutrophic reservoir in South Wales with emphasis on the Chironomidae (Diptera); their life-histories and production. *Arch. Hydrobiol.* **74**, 186–226.

Power, G. (1973). Estimates of age, growth, standing crop and production of salmonids in some North Norwegian rivers and streams. *Rep. Inst. Freshwat. Res. Drottningholm* **53**, 78–111.

Resh, V. H. (1975). The use of transect sampling in estimating single species production of aquatic insects. *Verh. Internat. Verein. Limnol.* **19**, 3089–3094.

Rey, J., and Capblancq, J. (1975). Dynamique des populations et production du zooplancton du Lac de Port-Bielh (Pyrénées Centrales). *Annls. Limnol.* **11**, 1–45.

Richman, S. (1971). Calorimetry. *In* "A Manual on Methods for the Assessment of Secondary Productivity in Fresh Waters" (Eds W. T. Edmondson and G. G. Winberg). IBP Handbook No. 17, Blackwell Sci. Publ., Oxford and Edinburgh, pp. 146–149.

Ricker, W. E. (1946). Production and utilization of fish populations. *Ecol. Monogr.* **16**, 373–391.

Ricker, W. E. (Ed.) (1968). "Methods for Assessment of Fish Production in Fresh Waters." IBP Handbook No. 3, Blackwell Sci. Publ., Oxford and Edinburgh.

Ricker, W. E. (1969). Effect of size-selective mortality and sampling bias on estimates of growth, mortality, production, and yield. *J. Fish. Res. Bd. Can.* **26**, 479–541.

Ricker, W. E. (1975). "Computation and Interpretation of Biological Statistics of Fish Populations." Bull. 191, Can. Dept. Environment, Fish., Mar. Serv., Ottawa.

Ricker, W. E. and Foerster, R. E. (1948). Computation of fish production. *Bull. Bingham Oceanogr. Coll., Yale Univ.,* **11** (art. 4), 173–211.

Rigler, F. H. and Cooley, J. M. (1974). The use of field data to derive population statistics of multivoltine copepods. *Limnol. Oceanogr.* **19**, 636–655.

Rigler, F. H., MacCallum, M. E. and Roff, J. C. (1974). Production of zooplankton in Char Lake. *J. Fish. Res. Bd. Can.* **31**, 637–646.

Russell-Hunter, W. D. (1970). "Aquatic Productivity." Macmillan, New York.

Salanki, J. and Ponyi, J. E. (Eds) (1975). "Limnology of Shallow Waters." Symp. Biol. Hungary 15, Akad. Kiado, Budapest.

Sanders, H. L. (1956). The biology of marine bottom communities. *Bull. Bingham Oceanogr. Coll., Yale Univ.* **15**, 345–414.

Saunders, L. H. and Power, G. (1970). Population ecology of the brook trout, *Salvelinus fontinalis*, in Matamek Lake, Quebec. *J. Fish. Res. Bd. Can.* **27**, 413–424.

Schindler, D. W. (1972). Production of phytoplankton and zooplankton in Canadian Shield lakes. *In* "Productivity Problems of Freshwaters" (Eds Z. Kajak and A. Hillbricht-Ilkowska). IBP, UNESCO, Polish Sci. Publ., Warsaw, pp. 311–331.

Schindler, D. W., Clark, A. S. and Gray, J. R. (1971). Seasonal calorific values of freshwater zooplankton, as determined with a Phillipson bomb calorimeter modified for small samples. *J. Fish. Res. Bd. Can.* **28**, 559–564.

Sheldon, A. L. (1972). Comparative ecology of *Arcynopteryx* and *Diura* (Plecoptera) in a California stream. *Arch. Hydrobiol.* **69**, 521–546.

Small, J. W., Jr. (1975). Energy dynamics of benthic fishes in a small Kentucky stream. *Ecology* **56**, 827–840.

Smyly, W. J. P. (1973). Bionomics of *Cyclops strenuus abyssorum* Sars (Copepoda: Cyclopoida). *Oecologia* **11**, 163–186.

Sokolowa, N. J. (1966). Biologie der Massenarten und Produktivität den Chironomiden in Utscha-Stausee. *Verh. Internat. Verein. Limnol.* **16**, 735–740.

Sorgeloos, P. and Persoone, G. (1973). A culture system for *Artemia, Daphnia,* and other invertebrates, with continuous separation of the larvae. *Arch. Hydrobiol.* **72**, 133–138.

Sorokin, Y. I. (1972). Biological productivity of the Rybinsk reservoir. *In* "Productivity Problems of Freshwaters" (Eds Z. Kajak and A. Hillbricht-Ilkowska). IBP, UNESCO, Polish Sci. Publ., Warsaw, pp. 493–503.

Sorokin, Y. I. and Kadota, H. (Eds) (1972). "Techniques for the Assessment of

Microbial Production and Decomposition in Fresh Waters." IBP Handbook No. 23, Blackwell Sci. Publ., Oxford and Edinburgh.

Speir, J. A. and Anderson, N. H. (1974). Use of emergence data for estimating annual production of aquatic insects. *Limnol. Oceanogr.* **19**, 154–156.

Staples, D. J. (1975). Production biology of the upland bully *Philypnodon breviceps* Stokell in a small New Zealand lake. III. Production, food consumption and efficiency of food utilization. *J. Fish Biol.* **7**, 47–69.

Stepanova, L. A. (1971). Production of some common plantonic crustaceans in Lake Ilmen. *Hydrobiol. J.* **7** (6), 13–23.

Stockner, J. G. (1971). Ecological energetics and natural history of *Hedriodiscus truquii* (Diptera) in two thermal spring communities. *J. Fish. Res. Bd. Can.* **28**, 73–94.

Stross, R. G., Neess, J. C. and Hasler, A. D. (1961). Turnover time and production of planktonic Crustacea in limed and reference portion of a bog lake. *Ecology* **42**, 237–245.

Teal, J. M. (1957). Community metabolism in a temperate cold spring. *Ecol. Monogr.* **27**, 283–302.

Tilley, L. J. (1968). The structure and dynamics of Cone Spring. *Ecol. Monogr.* **38**, 169–197.

Toetz, D. W. (1967). The importance of gamete losses in measurements of freshwater fish production. *Ecology* **48**, 1017–1020.

Tsuda, M. (1972). Interim results of the Yoshino River productivity survey, especially on benthic animals. *In* "Productivity Problems of Freshwaters" (Eds. Z. Kajak and A. Hillbricht-Ilkowska). IBP, UNESCO, Polish Sci. Publ., Warsaw, pp. 827–841.

Tudorancea, C. (1972). Studies on Unionidae populations from the Crapina-Jijila complex of pools (Danube zone liable to inundation). *Hydrobiologia* **39**, 527–561.

Vollenweider, R. A. (Ed.) (1969). "A Manual on Methods for Measuring Primary Production in Aquatic Environments." IBP Handbook No. 12, Blackwell Sci. Publ., Oxford and Edinburgh.

Ward, F. J. and Robinson, G. G. C. (1974). A review of research on the limnology of West Blue Lake, Manitoba. *J. Fish. Res. Bd. Can.* **31**, 977–1005.

Warren, C. E. (1971). "Biology and Water Pollution Control." Saunders, Philadelphia.

Warren, C. E., Wales, J. H., Davis, G. E. and Doudoroff, P. (1964). Trout production in an experimental stream enriched with sucrose. *J. Wildl. Mgmt.* **28**, 617–660.

Waters, T. F. (1962). A method to estimate the production rate of a stream bottom invertebrate. *Trans. Am. Fish. Soc.* **91**, 243–250.

Waters, T. F. (1966). Production rate, population density, and drift of a stream invertebrate. *Ecology* **47**, 595–604.

Waters, T. F. (1969). The turnover ratio in production ecology of freshwater invertebrates. *Am. Nat.* **103**, 173–185.

Waters, T. F. and Crawford, G. W. (1973). Annual production of a stream mayfly population: A comparison of methods. *Limnol. Oceanogr.* **18**, 286–296.

Welch, H. E. (1976). Ecology of Chironomidae (Diptera) in a polar lake. *J. Fish. Res. Bd. Can.* **33**, 227–247.

Welch, P. S. (1935). "Limnology." McGraw-Hill, New York and London.

Williams, D. D. and Hynes, H. B. N. (1974). The occurrence of benthos deep in the substratum of a stream. *Freshwat. Biol.* **4**, 233–256.

Winberg, G. G. (Ed.) (1971). "Methods for the Estimation of Production of Aquatic Animals." (Transl. by Annie Duncan, Academic Press, London.)

Winberg, G. G. (1972). Some interim results of Soviet IBP investigations on lakes. *In* "Productivity Problems of Freshwaters" (Eds Z. Kajak and A. Hillbricht-Ilkowska). IBP, UNESCO, Polish Sci. Publ., Warsaw, pp. 363–381.

Winberg, G. G. and Bauer, O. N. (1971). Productivity and principles of the management of inland waters in the USSR. *Freshwat. Biol.* **1**, 159–167.

Winberg, G. G., Pechen, G. A. and Shushkina, E. A. (1965). The production of planktonic crustaceans in three different types of lakes. *Zool. Zh.* **44**, 676–688. (In Russian, Transl. NLL RTS 6019).

Winberg, G. G., Babitsky, V. A., Gavrilov, S. I., Gladky, G. V., Zakharenkov, I. S., Kovalevskaya, R. Z., Mikheeva, T. M., Novyadomskaya, P. S., Ostapenya, A. P., Petrovich, P. G., Potaenko, J. S. and Yakushko, O. F. (1972). Biological productivity of different types of lakes. *In* "Productivity Problems of Freshwaters" (Eds Z. Kajak and A. Hillbricht-Ilkowska). IBP, UNESCO, Polish Sci. Publ., Warsaw, pp. 384–404.

Winberg, G. G., Alimov, A. F., Boullion, V. V., Ivanova, M. B., Korobtzova, E. V., Kuzmitzkaya, N. K., Nikulina, V. N., Finogenova, N. P. and Fursenko, M. V. (1973). Biological productivity of two subarctic lakes. *Freshwat. Biol.* **3**, 177–197.

Winterbourn, M. J. (1974). The life histories, trophic relations and production of *Stenoperla prasina* (Plecoptera) and *Deleatidium* sp. (Ephemeroptera) in a New Zealand river. *Freshwat. Biol.* **4**, 507–524.

Wright, J. C. (1965). The population dynamics and production of *Daphnia* in Canyon Ferry Reservoir, Montana. *Limnol. Oceanogr.* **10**, 583–590.

Yablonskaya, E. A. (1962). Study of the seasonal population dynamics of the plankton copepods as a method of the determination of their production. *Rapp. P.-V. Reun. Cons. Int. Explor. Mer* **153**, 224–226.

Yamamoto, G. (1972). Trophic structure in Lake Tatsu-Numa, an acidotrophic lake in Japan, with special reference to the importance of the terrestrial community. *In* "Productivity Problems of Freshwaters" (Eds Z. Kajak and A. Hillbricht-Ilkowska). IBP, UNESCO, Polish Sci. Publ., Warsaw, pp. 405–419.

Zaika, V. E. (1973). "Specific Production of Aquatic Invertebrates." Halsted Press, New York.

Zawislak, W. (1972). Production of crustacean zooplankton in Moty Bay, Lake Jeziorak. Part I. The method of production estimation. *Pol. Arch. Hydrobiol.* **19**, 179–191.

Zelinka, M. (1973). Die Eintagsfliegen (Ephemeroptera) in Forellenbächen der Beskiden. II. Produktion. *Hydrobiologia* **42**, 13–19.

Zhdanova, G. A. (1969). Comparative characteristics of the life cycle and productivity of *Bosmina longirostris* O. F. Müller and *B. coregoni* Baird in the Kiev Reservoir. *Hydrobiol. J.* **5** (1), 8–15.

Zwick, P. (1975). Critical notes on a proposed method to estimate production. *Freshwat. Biol.* **5**, 65–70.

Author Index

M

Maiorana, V. C., 61, *62*

Maitland, P. S., 102, 126, 127, 128, 136, 137, *159*

Mallach, N., 66, *88*

Mann, K. H., 93, 100, 102, 111, 113, 116, 119, 124, 131, 132, 133, 134, 135, 137, 138, 140, 148, 149, *152*, *159*

Mann, R. H. K., 117, 120, 122, 123, 126, 148, *153*, *154*, *159*

Markus, H. C., 40, *58*

Marples, T. G., 20, *59*

Marsh, R. E., 80, *87*

Marshlll, S. G., 20, *59*

Marzolf, G. R., 104, 129, *156*

Mason, J. W., 122, 148, *152*

Mathews, C. P., 99, 101, 112, 116, 118, 119, 120, 125, 138, 150, *159*, *160*

Mathias, J. A., 132, *160*

Maynard Smith, J., 40, *58*

Mayr, E., 3, 16, *59*

Medawar, P. B., 50, *59*

Mendel, L. B., 30, *59*

Metzgar, L. H., 71, 78, *88*

Michailova, N. F., 138, 145, *151*

Mikheeva, T. M., 119, 139, 142, *164*

Miller, R. B., 98, *160*

Mills, D. H., 120, *160*

Miner, R. W., 36, *59*

Mittelholzer, E., 143, 144, 146, *160*

Momot, W. T., 112, 132, 133, 137, 140, *160*

Moody, P. A., 81, *88*

Morgan, N. C., 128, 136, *159*

Mori, S., 94, *160*

Morris, R. D., 75, 77, *88*

Moshiro, G. A., 116, 141, 143, 144, *154*, *160*

Moskalenko, B. K., 141, 143, 146, *160*

Mullin, M. M., 116, *160*

Murdoch, W. W., 40, *59*

Murie, A., 66, 70, 71, 73, 77, *88*

Murie, M., 65, 66, 68, 71, 72, 73, 74, 75, 76, 77, 83, *88*

Murie, O. J., 66, 70, 71, 73, 77, *88*

N

National Academy of Sciences, 94, *160*

Nechvalenko, S. P., 99, *158*

Needham, A. E., 44, *59*

Neess, J., 101, *160*

Neess, J. C., 107, 110, 144, 149, *163*

Negus, C., 134, 140, *160*

Nelson, D. J., 138, *160*

Neveu, A., 102, 104, 129, 137, *160*

Nevo, E., 16, *59*

Nikulina, V. N., 114, 138, 139, 141, 142, 145, 146, 147, *151*, *164*

Novyadomskaya, P. S., 119, 139, 142, *164*

O

Ockelman, K. W., 24, *55*

O'Connor, J. F., 117, 120, 122, 126, *160*

Odum, E. P., 18, 20, *59*, 96, *161*

Odum, H. T., 9, 10, *59*, 98, *161*

Orgel, L. E., 51, *59*

Orr, R., 64, *88*

Osbourn, D. F., 30, *61*

Osbourne, T. B., 30, *59*

Ostapenya, A. P., 119, 139, 142, *164*

Otto, C., 114, 131, 137, *161*

Ozertskovskaya, N. G., 138, 145, *151*

P

Paine, R. T., 37, *59*, 116, *161*

Paloheimo, J. E., 30, *59*, 107, *161*

Pandian, T. J., 28, *59*

Parnas, H., 20, *59*

Parsons, L., 79, 80, *88*

Patalas, K., 148, *161*

Pearson, W. D., 130, 131, *161*

Pechen, G. A., 99, 107, 110, *161*, *164*

Pentelow, F. J. K., 28, *59*

Persoone, G., 106, *162*

Peterson, B., 18, *59*

Petrosky, C. E., 106, 117, 125, 148, *161*

Petrovich, P. G., 119, 139, 142, *164*

Petrusewicz, K., 94, *161*

Pianka, E. R., 61, *62*

Pidgaiko, M. L., 138, 148, *161*

Pieczynska, E., 138, 141, 145, 147, *157*

Pinkerton, R., 9, 10, *59*

Pivnicka, K., 119, 120, *156*

Platt, T., 116, *161*

Polivannaya, M. F., 138, 148, *161*

Ponyi, J. E., 94, *162*

Potaenko, J. S., 119, 139, 142, *164*

Subject Index

Advances in Ecological Research, Volumes 1—9: Cumulative List of Titles

Analysis of processes involved in the natural control of insects, **2**, 1

The distribution and abundance of lake-dwelling Triclads—towards a hypothesis, **3**, 1

The dynamics of aquatic ecosystems, **6**, 1

The dynamics of a field population of the pine looper, *Bupalus piniarius* L. (Lep., Geom.), **3**, 207

Ecological aspects of fishery research, **7**, 115

Ecological conditions affecting the production of wild herbivorous mammals on grasslands, **6**, 137

Ecological implications of dividing plants into groups with distinct photosynthetic production capacities, **7**, 87

Ecological studies at Lough Ine, **4**, 198

Ecology of fire in grasslands, **5**, 209

The ecology of serpentine soils, **9**, 255

Ecology, systematics and evolution of Australian frogs, **5**, 37

Energetics, terrestrial field studies, and animal productivity, **3**, 73

Energy in animal ecology, **1**, 69

Forty years of genecology, **2**, 159

The general biology and thermal balance of penguins, **4**, 131

Heavy metal tolerance in plants, **7**, 2

Human ecology as an interdisciplinary concept: a critical inquiry, **8**, 2

Integration, identity and stability in the plant association, **6**, 84

Litter production in forests of the world, **2**, 101

Mathematical model building with an application to determine the distribution of Dursban® insecticide added to a simulated ecosystem, **9**, 133

The method of successive approximation in descriptive ecology, **1**, 35

Pattern and process in competition, **4**, 1

Population cycles in small mammals, **8**, 268

Predation and population stability, **9**, 1

The pressure chamber as an instrument for ecological research, **9**, 165

The production of marine plankton, **3**, 117

Quantitative ecology and the woodland ecosystem concept, **1**, 103

Realistic models in population ecology, **8**, 200

A simulation model of animal movement patterns, **6**, 185

Studies on the cereal ecosystem, **8**, 108

Studies on the insect fauna on Scotch Broom *Sarothamnus scoparius* (L.) Wimmer, **5**, 88

Soil arthropod sampling, **1**, 1

A synopsis of the pesticide problem, **4**, 75

Towards understanding ecosystems, **5**, 1

The use of statistics in phytosociology, **2**, 59

Vegetational distribution, tree growth and crop success in relation to recent climatic change, **7**, 177